Mechanical Behavior of Biomaterials

Woodhead Publishing Series in Biomaterials

Mechanical Behavior of Biomaterials

Edited by

J. Paulo Davim

Department of Mechanical Engineering,
University of Aveiro, Aveiro, Portugal

WP
WOODHEAD PUBLISHING
ELSEVIER An imprint of Elsevier

Woodhead Publishing is an imprint of Elsevier
The Officers' Mess Business Centre, Royston Road, Duxford, CB22 4QH, United Kingdom
50 Hampshire Street, 5th Floor, Cambridge, MA 02139, United States
The Boulevard, Langford Lane, Kidlington, OX5 1GB, United Kingdom

Library of Congress Cataloging-in-Publication Data
A catalog record for this book is available from the Library of Congress

British Library Cataloguing-in-Publication Data
A catalogue record for this book is available from the British Library

ISBN: 978-0-08-102174-3 (print)
ISBN: 978-0-08-102175-0 (online)

For information on all Woodhead publications visit
our website at https://www.elsevier.com/books-and-journals

Publisher: Matthew Deans
Acquisition Editor: Laura Overend
Editorial Project Manager: Lindsay Lawrence
Production Project Manager: Maria Bernard
Cover Designer: Miles Hitchen

Typeset by SPi Global, India

Contents

Contributors

Iswadi Ahmad Kolej Kemahiran Tinggi MARA Kuantan, Kuantan, Malaysia

A.K. Basak Adelaide microscopy, The University of Adelaide, Adelaide, SA, Australia

L.C. Brandão Department of Mechanical Engineering, Centre for Innovation in Sustainable Manufacturing, Federal University of São João del Rei, São João del Rei, Brazil

D. Carou Department of Mechanical Engineering, University of Aveiro, Aveiro, Portugal; Department of Mechanical and Mining Engineering, University of Jaén, Jaén, Spain

S. Chattopadhyaya Department of Mech. Eng., IIT-ISM, Dhanbad, India

Suman Kalyan Das Jadavpur University, Kolkata, India

J. Paulo Davim Department of Mechanical Engineering, University of Aveiro, Aveiro, Portugal

Linxia Gu Department of Mechanical and Materials Engineering, University of Nebraska-Lincoln, Lincoln, NE, United States

Wan Sharuzi Wan Harun Faculty of Mechanical & Manufacturing Engineering, Universiti Malaysia Pahang, Pekan, Malaysia

Luke Hocking School of Civil and Mechanical Engineering, Curtin University, Bentley, WA, Australia

Kumaran Kadirgama Faculty of Mechanical & Manufacturing Engineering, Universiti Malaysia Pahang, Pekan, Malaysia

Kaushik Kumar Department of Mechanical Engineering, BIT Mesra, Ranchi, India

C.H. Lauro Department of Mechanical Engineering, University of Aveiro, Aveiro, Portugal; Department of Mechanical Engineering, Centre for Innovation in Sustainable Manufacturing, Federal University of São João del Rei, São João del Rei, Brazil

Mahmoud Moradi Department of Mechanical Engineering, Malayer University, Malayer, Iran

J.C. Nogueira Department of Mechanical Engineering, University of Aveiro, Aveiro, Portugal

A. Pramanik School of Civil and Mechanical Engineering, Curtin University, Bentley, WA, Australia

Devarajan Ramasamy Faculty of Mechanical & Manufacturing Engineering, Universiti Malaysia Pahang, Pekan, Malaysia

Mahendran Samykano Faculty of Mechanical & Manufacturing Engineering, Universiti Malaysia Pahang, Pekan, Malaysia

Prasanta Sahoo Jadavpur University, Kolkata, India

Junfei Tong Department of Mechanical and Materials Engineering, University of Nebraska-Lincoln, Lincoln, NE, United States

Divya Zindani Department of Mechanical Engineering, NIT Silchar, Silchar, India

About the editor

J. Paulo Davim received his PhD degree in mechanical engineering in 1997, his MSc degree in mechanical engineering (materials and manufacturing processes) in 1991, and his mechanical engineering degree (5 years) in 1986, from the University of Porto (FEUP); the Aggregate title (Full Habilitation) from the University of Coimbra in 2005; and his DSc from London Metropolitan University in 2013. He is a Senior Chartered Engineer by the Portuguese Institution of Engineers, with an MBA and Specialist title in engineering and industrial management. He is also Eur Ing by FEANI-Brussels and Fellow (FIET) by IET-London. Currently, he is professor at the Department of Mechanical Engineering of the University of Aveiro, Portugal. He has more than 30 years of teaching and research experience in manufacturing, materials, mechanical and industrial Engineering, with special emphasis in machining and tribology. He also has an interest in management, engineering education, and higher education for sustainability. He has guided large numbers of postdoctoral, PhD, and master's degree students, as well as coordinated and participated in several financed research projects. He has received several scientific awards. He has worked as evaluator of projects for the European Research Council (ERC) and other international research agencies as well as examiner of PhD theses for many universities in different countries. He is the editor in chief of several international journals, guest editor of many journals, book editor, book series editor, and scientific advisor for many international journals and conferences. Presently, he is an editorial board member of 30 international journals and acts as reviewer for more than 100 prestigious Web of Science journals. In addition, he has also published as editor (and coeditor) more than 100 books and as author (and coauthor) more than 10 books, 80 book chapters, and 400 articles in journals and conferences (more than 250 articles in journals indexed in Web of Science core collection/h-index 50+/7500+ citations, SCOPUS/h-index 56+/10500+ citations, Google Scholar/h-index 71+/16500+ citations).

Preface

According to the US National Institutes of Health, biomaterials can be defined as "*any substance or combination of substances, other than drugs, synthetic or natural in origin, which can be used for any period of time, which augments or replaces partially or totally any tissue, organ or function of the body, in order to maintain or improve the quality of life of the individual.*" This book considers aspects of biomaterials related to biotribology, mechanical deformation, damage, failure under applied forces, and manufacturing. The book aims to provide the latest information on the mechanical behavior of biomaterials for the modern biomedical industry.

The initial chapter of the book discusses the tribology of materials for biomedical applications. Chapter 2 covers designing and analysis of the femoral neck for an artificial hip joint prosthesis. Chapter 3 presents biomechanics of the optic nerve head. Chapter 4 covers metallic biomaterials (a review). Chapter 5 is dedicated to the mechanical behavior of selective laser melting–produced metallic biomaterials. The last chapter of the book is devoted to the influence of dry and near-dry machining in a biomaterial with dual negative tool geometry.

This work is suitable as a text for a senior undergraduate engineering course or as a topic on mechanical behavior of biomaterials at the postgraduate level. It can also serve as a valuable reference for academics, researchers, medical doctors, and biomedical, mechanical and materials engineers, as well as professionals working in the biomedical industry. This material is of scientific interest to institutes, research centers, laboratories, and universities worldwide.

The editor acknowledges Elsevier for this opportunity and for their professional support. Finally, I would like to thank all the chapter authors for making themselves available to contribute to this work.

J. Paulo Davim
Aveiro, Portugal

Tribology of materials for biomedical applications

1

Prasanta Sahoo, Suman Kalyan Das*, J. Paulo Davim†*
*Jadavpur University, Kolkata, India, †Department of Mechanical Engineering, University of Aveiro, Aveiro, Portugal

1.1 Introduction

A long healthy life—in recent years we have come a long way towards fulfilling this age-old dream of mankind, thanks to advancements in modern medicine. But this very success creates a host of new challenges for medicine. Our increasingly aging population has produced a rise in age-related ailments. Our diet is different from that of our grandparents, resulting in obesity and metabolic disorders. Also, the trend towards high-risk recreational sports persists, with potential hazards ranging from fractures to severe internal injuries. Present-day and future medicine must confront these changes in modern society.

Since ancient times, humans have attempted to restore the functionalities of body parts stricken with trauma or disease. The human body and its associated biological systems are unique. Commonly available materials in their raw forms, when directly interacting with these biological systems, may result in various side effects and damage to the human body. Hence, some special materials have been identified, called biomaterials, that are both compatible with living tissue and provide the necessary engineering functions. Metals and alloys, ceramics, and polymer-based materials are often used in implants and other medical devices. Fig. 1.1 illustrates some of the metallic implants and bone fixation devices available.

The science of tribology is not limited to mechanical machinery; it also finds application in the medical field. The human body possesses a wide variety of sliding and frictional interfaces, mainly in the joints. Moreover, the friction between the eyelids and eyeball, skin friction, etc. also fall under the scope of tribology. Hence, a separate domain, called biotribology, has been developed to deal with the application of tribological principles, such as friction, wear, and lubrication between interacting surfaces in relative motion, to medical and biological systems [2]. According to www.nature.com, "Biomedical materials are biomaterials that are manufactured or processed to be suitable for use as medical devices (or components thereof) and that are usually intended to be in long-term contact with biological materials." Study of the tribological aspects of biomedical materials is equally important, as such study deals with reducing friction and wear, thus resulting in the greater longevity of biomedical implants and devices. This reduces the complications associated with repeated surgeries. Besides, as more cases of younger implant patients are appearing, increasing the longevity of implants has become very significant. Typical examples of tribology in biomedicine include the following:

- Tribology of natural synovial joints and artificial replacements
- Wear of dental implants

Mechanical Behavior of Biomaterials. https://doi.org/10.1016/B978-0-08-102174-3.00001-2

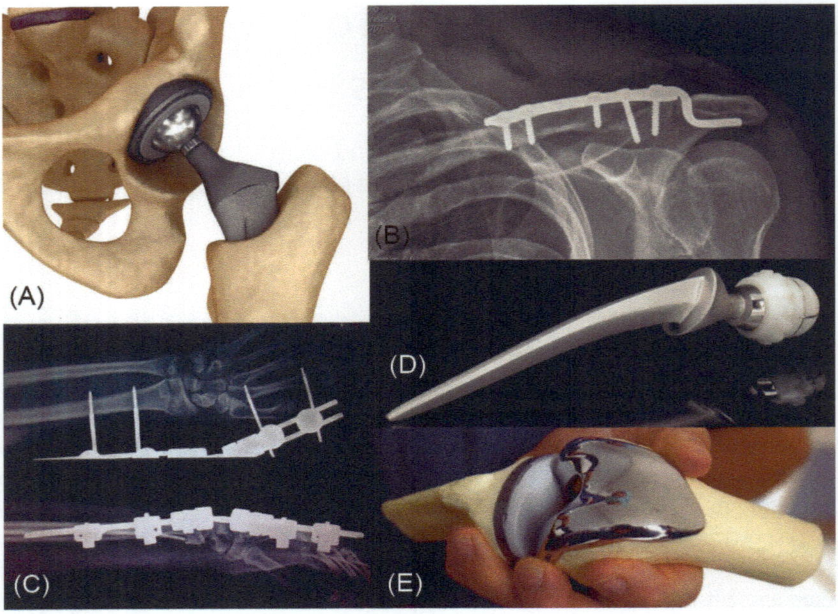

Fig. 1.1 Examples of metallic implants and bone fixation devices [1].
(A) From https://www.lawyersandsettlements.com/images/articles2/hip17-article.jpg.
(B) From https://upload.wikimedia.org/wikipedia/commons/a/a9/Claviculafraktur_lateral_6mo_platte.jpg. (C) From https://upload.wikimedia.org/wikipedia/commons/3/3c/External_fixator_xray.jpg. (D) From https://upload.wikimedia.org/wikipedia/commons/0/00/Stainless_steel_and_ultra_high_molecular_weight_polythene_hip_replacement_(9672239334).jpg. (E) From http://www.iran-daily.com/content/imgcache/file/132951/0/image_650_365.jpg.

- Wear and replacement of heart valves
- Lubrication of pump in total artificial hearts
- Ocular tribology and tribology of contact lenses
- Wear of screws and plates in bone fracture repair
- Friction of skin and interaction with clothing

Finally, with growing knowledge of the tribological aspects of biomedicine, the quality of human life is expected to improve.

1.2 Desired properties in biomaterials for medical applications

Materials to be used within the living human body, and supposed to coexist with living tissue and other organic matter without any degradation, should possess unique combinations of properties, some of which are given in the following list [3]:

- *Biocompatibility:* The biomaterial should be compatible with living systems and not cause any bodily harm, which includes any negative effects a material can have on the components of a biological system (bone, extra- and intracellular tissues, and ionic composition of plasma).
- *Nontoxic:* The material should not be toxic to living cells and organisms. Toxicity can be of two types: genotoxic (which can alter the DNA of the genome) or cytotoxic (causes damage to individual cells). Failure to comply with both biocompatibility as well as nontoxicity can lead to rejection of implants and other serious health conditions.
- *Mechanical properties:* The material should have a low modulus combined with high strength to prolong the service period of the implant and prevent loosening, thereby preventing the need for revision surgery. Moreover, stress shielding (reduction in bone density as a result of removal of typical stress from the bone by an implant) can be prevented by matching the modulus of elasticity of biomaterials to that of bone, which varies from 4 to 30 GPa.
- *High wear resistance:* The material should have a high wear resistance and exhibit a low friction coefficient when sliding against body tissues. An increase in the friction coefficient or a decrease in the wear resistance can cause the implant to loosen. Moreover, the wear debris generated can cause inflammation destructive to the bone supporting the implant.
- *High corrosion resistance:* The human body is not an environment that one would consider hospitable for an implanted metal alloy: a highly oxygenated saline electrolyte at a pH of around 7.4 and a temperature of 37°C [4]. Moreover, the abundant presence of chlorine ions in the body fluids results in aggravated corrosion scenarios for metals. An implant made of a biomaterial with a low corrosion resistance can release metal ions into the body, which in turn produces toxic reactions. Thus, high corrosion resistance is a desired characteristic of biomaterials.
- *Long fatigue life:* The joints in a human body are subjected to cyclic motion as well as cyclic variation in loading throughout a person's life. Hence, the material should exhibit a high resistance to failure by fatigue to prevent implant failure and stress shielding from fatigue fracture. The failure of implants by fatigue has been reported for hip prostheses.
- *Osseointegration:* Osseointegration was first defined as "a direct structural and functional connection between ordered, living bone and the surface of a load-carrying implant." The roughness, chemistry, and topography of the surface play a major role in good osseointegration. Implant loosening results from the nonintegration of the implant surface into the adjacent bone. A few researchers mention that osseointegration is undesirable due to the risk of not being able to remove the implant after use. However, a few of them have also demonstrated that the implant can be removed safely. Thus osseointegration is a desirable property for a biomaterial in some applications, such as in an implant, where it must be ensured that the implant will integrate properly with the bone and other tissues.

1.3 What is tribology?

Tribology is the study of the phenomena related to the surface of a solid or the interface between two surfaces. Friction, wear, roughness, etc. are the tribological characteristics. Friction is the force resisting the relative motion between two surfaces sliding against each other, whereas wear is loss of material or deformation of a body when two surfaces in contact have relative motion between them. Among the two degrading phenomena, friction dictates the efficiency of mechanical assemblies that involve sliding surface contact. It is also responsible for wear, which is often the limiting mechanism of device service life. Thus, minimization of friction and wear is vital where high efficiency and longer device life are needed. Recently, corrosion has been closely related to tribology as it results in surface degradation that promotes wear. Due to this synergistic effect between corrosion and wear, the term *tribocorrosion* has been coined, which takes into account the combined effect of corrosion and wear. Tribocorrosion is found to occur in many engineering fields. Components like pipes, valves, pumps, waste incinerators, mining equipment, medical implants, etc. are subjected to a tribocorrosion effect when in operation, which can reduce their lifetime. Besides, the safety of critical systems like nuclear reactors and human transportation systems is threatened due to tribocorrosion.

The study of the tribological aspect of systems has tremendous economic and technological importance. Losses due to energy dissipation and material wear add up to billions of dollars annually in industrialized countries. A proper understanding of tribological processes can provide a basis on which to improve standards of design and increase engineering efficiency. This has already had a significant impact on energy conservation issues that concern the future of mankind.

1.3.1 About biotribology

Although tribology has conventionally been associated with the surface interaction of mechanical systems, concepts of tribology have also been important in the study of biological systems. Biotribology is one of the newest fields to emerge in the discipline of tribology. Biotribology deals with all aspects of tribology concerned with biological systems [5]. It is recognized as one of the most important considerations in many biological systems, contributing to the understanding of how our natural systems work. In addition, it helps in understanding how diseases develop and how medical interventions should be applied.

Biotribology is one of the most exciting areas of tribology research and is one that affects various aspects of our everyday lives, from skin blisters to artificial joints and contact lenses. Categorical research works on biotribology are presented in Table 1.1. In many cases interaction with our environment is governed by tribology, and in particular our response to perceived friction. The use of touch to evaluate surface texture hydration and grip is an important example. In addition to those disciplines associated with tribology, biotribology also involves biomechanics, biochemistry, biology, physiology, clinical medicine, and pathology. Increasingly, biotribology research is contributing significant scientific, social, and healthcare benefits; the opportunities are considerable [6].

1.4 Biomedical engineering applications

Biomedical engineering is the application of engineering principles and design concepts to medicine and biology for healthcare purposes (e.g., diagnostic or therapeutic). Biomedical engineers work at the intersection of engineering principles, the life sciences, and healthcare. These engineers take principles from applied science (including mechanical, electrical, chemical, and computer engineering) and physical sciences (including physics, chemistry, and mathematics) and apply them to biology and medicine.

Table 1.1 Classification of biotribology research and associated areas [5]

Classification type	Major areas of investigation
Natural joint	Synovial joints, articular cartilage, meniscus, mechanically and biochemically induced damage, etc.
Artificial articular joint	Partial and total joint replacement (hip and knee), spinal discs, explant analysis, implant corrosion and wear, artificial cartilage, bioscaffolds, etc.
Oral tribology	Natural teeth, tongue, mandibular joints, saliva, implant teeth, toothpaste, swallow, dental restorative materials, etc.
Skin tribology	Skin friction-induced perception; skin care; synthetic skin; skin in contact with articles (such as tactile texture, shaving devices, shoes, socks) for daily use, various medical as well as sport devices, medical and cosmetic treatment; skin friction and grip of objects; skin irritation and discomfort; etc.
Ocular tribology	Ocular surfaces, contact lenses, tear lubrication and dry eye syndrome, etc.
Prosthesis tribology	Prosthetic human interfacing and coupling, tribological function, etc.
Haptics	Tactile perception and surface texture, ergonomics, etc.
Medical devices	Scalpel, operation forceps, urinary catheters, gastroscope, artificial cardiovascular system, medical gloves, etc.
Sports tribology	Equipment design and development, preparation, deterioration and testing of sport surfaces, grip, player interaction and gait analysis, etc.
Biomimetics	Bioinspired tribology, insect tribology, etc.

Although the human body is a more complex system than even the most sophisticated of machines, many of the same concepts that go into building and programming a machine can be applied to biological structures and diagnostic and therapeutic tools.

Prominent biomedical engineering applications include the development of biocompatible prostheses and various diagnostic and therapeutic medical devices, ranging from clinical equipment to microimplants, common imaging equipment such as magnetic resonance imaging (MRI) and electroencephalography (EEG), tissue and stem cell engineering, clinical engineering, pharmaceutical drugs, and therapeutic biologicals.

1.4.1 Tribological links in biomedical applications

As already discussed, tribology has evolved as an important field both in conventional engineering as well as in the medical domain. The study of tribological challenges in the medical domain is one of the newest fields of study to have emerged in the area of tribology. Biomedical applications such as arthroplasty (especially in knee and hip joints), artificial hearts, dental implants, etc. encounter friction and wear between surfaces and hence present perfect cases for the incorporation of tribological knowledge for their fine tuning. Fig. 1.2 shows how tribology influences the biomedical area.

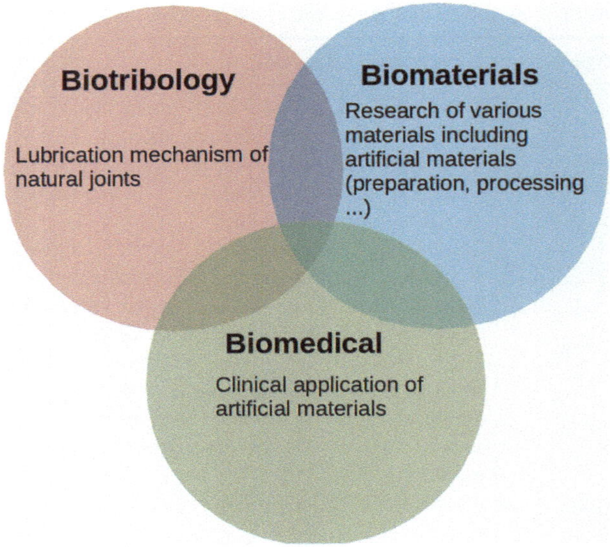

Fig. 1.2 Influence of tribology in biomedical engineering.
From https://commons.wikimedia.org/wiki/File:Biotribology.jpeg#filehistory.

Traditionally, biomechanical studies of joints are based on the quest to better understand their structure-function relationship in providing joint motion and the pathomechanical processes involved in joint diseases such as osteoarthritis [7]. The target of biotribologists is to evaluate biological systems and understand how they function with such tribological efficiency, providing increased understanding of their normal, as well as their pathologic, states. In a synovial joint, the synovial fluid, articular cartilage, and the supporting bone form a bearing system. The performance of such a joint depends on the mechanical behavior of the materials that make up the joint. Imagining a joint disease as the failure of bearing lubrication processes is an obvious oversimplification. However, the correlation between an engineering bearing and a synovial joint is an appropriate one. Examples of tribology in biomedicine include the study of lubrication by synovial fluid, measurement of friction in synovial joints, the mechanisms of joint lubrication, measurement and analysis of cartilage wear and damage, study of joint mechanics, and the development of artificial joints. In addition, the tribology of dental implants and ocular tribology also fall under the purview of this domain. Research in this field has led to the development of newer materials suitable for biomedical applications. This has given relief to thousands of diseased persons worldwide, thus benefiting society as a whole.

1.5 Artificial joints: Arthroplasty

One of the successfully implemented medical devices in the human body is the artificial joint. The surgical reconstruction or replacement of a joint is termed arthroplasty.

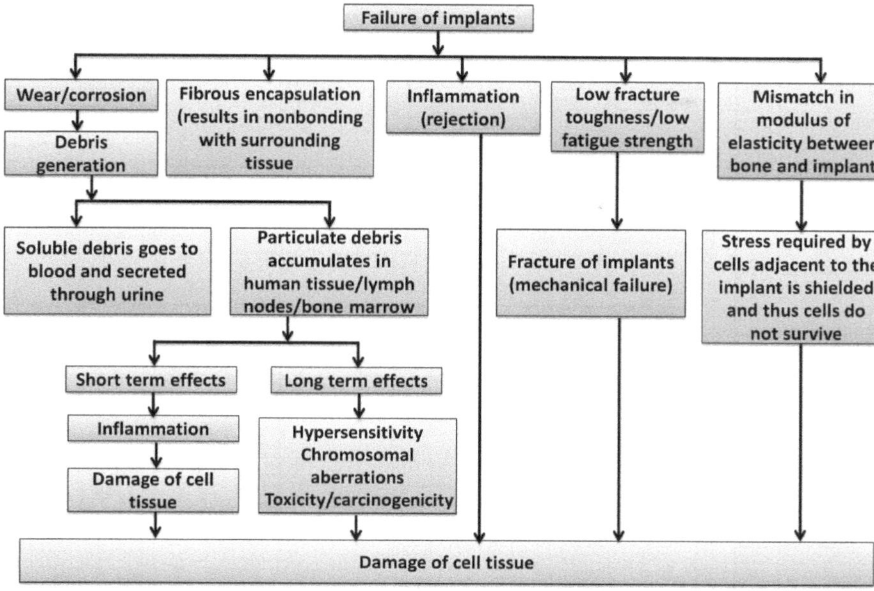

Fig. 1.3 Reported causes of implant failure [8].

There are 206 bones and over 300 joints in the human body [2]. Of these, the joints that allow larger relative motion are the hip joint, the knee joint, the shoulder joint, and the neck joint. Smaller joint implants such as the ankle, the elbow, the wrist, and the finger are also increasingly being introduced into medical practices. Arthroplasty is also considered for spinal disc (total disc) replacement and the temporomandibular joint (TMJ) prosthesis. Complex three-dimensional motion is experienced in these joints and that too is under a significant amount of loading. Reported causes of implant failure are illustrated in Fig. 1.3. Tribological issues at these joints, particularly the articulating surfaces, warrant careful consideration. Friction, wear, lubrication, and sometimes corrosion play important roles in the successful function of artificial joints.

1.5.1 Types of articulating surface

Articulating surfaces can broadly be divided into soft-on-hard and hard-on-hard combinations.

- Soft-on-hard combinations mainly include the following material combinations: ultra-high molecular weight polyethylene (UHMWPE) against cobalt chromium alloys or alumina/zirconia toughened alumina composite ceramics (ZTA). Titanium alloys are sometimes preferred, particularly for total disc replacements in the spine, due to their lower elastic modules and improved imaging quality, but surface treatments to improve wear resistance are necessary.
- Hard-on-hard bearing surface combinations for a hip implant include the following material combinations: metal-on-metal, ceramic-on-ceramic and ceramic-on-metal.

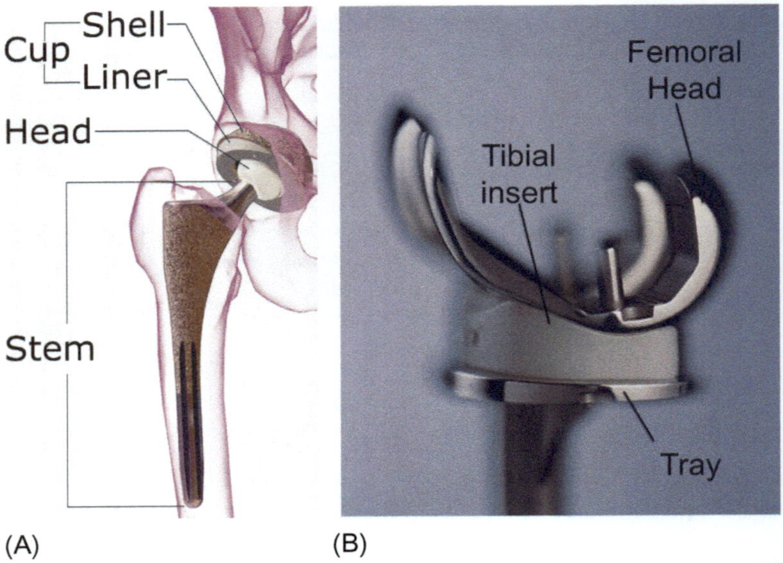

Fig. 1.4 Typical hip and knee implant with components. (A) Hip implant: metallic femoral head, plastic cup, and metallic backing shell; and (B) Knee: metallic femoral head, plastic tibial insert, and metallic tray.
(A) From https://commons.wikimedia.org/wiki/File:Hip_prosthesis_components.jpg. (B) Used with permission from DePuy Synthes.

The components of artificial hip and knee implants are illustrated in Fig. 1.4. The major tribological issue in artificial joints is wear and the resulting wear debris, which can be the cause of adverse tissue reactions and infections that may lead to loosening of the prosthetic components. Hence, there has been a major effort to increase the wear resistance of the bearing surface.

1.5.2 Biological reactions to wear debris in joint replacements

The majority of the arthroplasties carried out today consist of a combination of hard metal or ceramic that articulates against polyethylene parts. In the case of total hip prostheses, the femoral head is made of metal/ceramic whereas the acetabular cup is made of UHMWPE. Like other joints with a large range of motion being submitted to cyclic and heavy loading, friction of the bearing surfaces produces wear particles that are susceptible to diffusing in the surrounding soft tissues [9] and also to migrating towards more remote organs [10].

Most of the total hip prostheses comprise a hard metal or ceramic femoral head articulating against a UHMWPE acetabular cup. Evidence over the years has shown that these prostheses are prone to failure due to aseptic loosening. Aseptic loosening occurs due to physical problems of the implant system and not due to any type of infection, or under the influence of any harmful bacteria, viruses, or other microorganisms. Due to these, very few implants are found to survive beyond 25 years [11]. With implants becoming more common in younger and more active patients, the need to understand the mechanisms of failure and to develop artificial hip joints using alternative materials have become major issues in the orthopedic community. The principal reason for the aseptic loosening of the prosthetic implant is the generation of wear debris as a result of biological reactions. Beyond a degree of wear, the fixation of the joint into the bone fails and the joints loses its functionality, requiring a revision surgery. Aseptic loosening may also occur due to the biological response of the bone to stress shielding, micromotion at the bone-cement and cement-prosthesis or bone-prosthesis interfaces. Moreover, biological loosening may also occur due to osteolysis caused by adverse cellular reactions to debris generated by wear.

According to Ingham and Fisher [11], UHMWPE wear debris generated at the articulating surfaces enters the periprosthetic tissue, where it is phagocytosed by macrophages. Pro-inflammatory cytokines and other mediators of inflammation released by macrophages then stimulate osteoclastic bone resorption, which leads to osteolysis and finally loosening of the prosthesis. This is evident through the analysis of retrieved tissues and in vitro and in vivo studies of the biological effects of wear debris. For this reason, there is an increased interest in the development and use of alternative bearing surfaces, viz. metal-on-metal and ceramic-on-ceramic, for artificial hip joints.

Scanning electron microscopy (SEM) is employed to determine the size of the debris, which is found to lie in the range 0.3–0.5 μm. McKellop et al. [12] estimated that hundreds of billions of particles would be generated each year if the UHMWPE cup had a wear volume of 40 mm^3/year. Wear volumes generated by metal-on-metal articulations have been shown to be 40–100 times lower than those generated by metal-on-polyethylene bearings [13]. Studies on metal-on-metal bearings show that the wear rate is strongly dependent on materials, tribological design, and surface finishing technique [14].

Ingham and Fischer [11] observed that it is not the volumetric wear of the prosthesis that is important. None of the modern prostheses will actually wear out. It is the number of particles generated within the biologically active size range, for a given wear volume, and the particle load within an area of tissue that will determine the duration of survival of the implant for any given individual. New generation metal-to-metal or ceramic-to-ceramic may provide the solution to late aseptic loosening.

1.6 Materials for implants

Biological factors together with the prosthetic design features influence the performance of the total joint prosthesis. The interface between the implant and surrounding tissues is influenced by the size and shape, materials, and surface characteristics of the implant [15]. Biocompatibility is a prime requisite for materials to be used as orthopedic and other implants. Biocompatibility implies that the implant does not interact adversely with the physiological environment or vice versa. A stable interface between the prosthesis and the surrounding tissues is necessary for a successful long-term fixation of the implant. The application of load through a prosthesis as well as the muscle force results in stresses and strains at the implant interface.

1.6.1 Choice of material for joint replacements

Two main concerns raised by researchers are:

(i) Searching alternative materials for joint replacement.
(ii) How can these materials be fixed firmly?

The first concern is related to bearings and researchers have surveyed a variety of materials including metals and alloys (e.g., Co-Cr), synthetic substances, polyethylene, and ceramics. The limitations of glass, ivory, and nylon have curbed their use, while polyethylene and ceramic materials are widely used because of their advantages. The second concern is related to the aseptic loosening of the implants due to wear and tear. For this, a thorough knowledge of the tribological behavior of the implant material is useful.

1.6.1.1 Materials for knee arthroplasty

An artificial knee joint comprises a flat metal plate with stem implanted in the tibia or shin bone, a polyethylene bearing surface, and a contoured metal implant that fits around the end of the femur, as shown in Fig. 1.5. The use of a soft-on-hard combination, e.g., metal and polyethylene, allows optimum joint mobility with negligible wear. As the bearing used in a knee joint is almost flat, wear is of a lesser concern compared

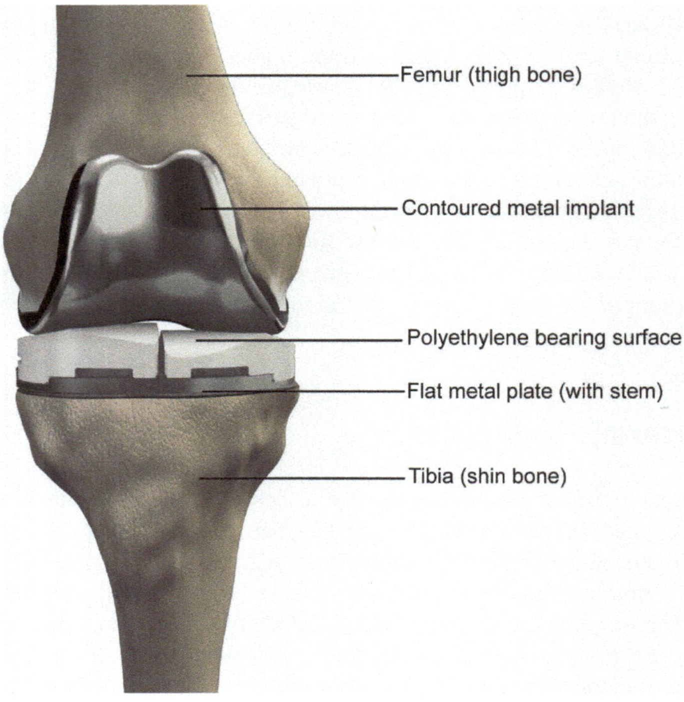

Fig. 1.5 Artificial knee joint.
From https://commons.wikimedia.org/wiki/File:Conformis_Kniegelenksprothese.jpg.

to hip implants, which have a very poor bearing capability. Materials used in knee implants are listed as follows (refer to https://bonesmart.org/).

Stainless steel

The use of stainless steel as an implant material was seen during the initial stages of the development of the field of prosthetics. However, due to the limited ability to withstand corrosion in the human body over the long term, the use of stainless steel as knee replacement implants was restricted. Stainless steel is more suited to be used as a temporary implant, such as fracture plates and screws.

Cobalt-chromium alloys

Cobalt-chromium (Co-Cr) alloys have high specific strength and are hard, tough, corrosion resistant, biocompatible metals. Besides titanium, cobalt chrome is one of the most widely used metals in knee implants. The good mechanical properties of Co-Cr alloys, similar to stainless steel, are due to the multiphase structure and precipitation of carbides, which increase the hardness of Co-Cr alloys enormously. The hardness of Co-Cr alloys ranges between 550 and 800 MPa, and tensile strength ranges between 145 and 270 MPa. Moreover, the tensile and fatigue strength increases radically as they are heat-treated. These alloys have properties quite similar to stainless steel.

Although the percentage of patients having allergic reactions related to the use of cobalt-chromium alloys is very low, one area of concern is the issue of tiny particles (metal ions) that may be released into the body as a result of joint movement. These particles can sometimes cause reactions in the human body, especially in the case of those patients who have allergies to particular metals like nickel.

Titanium and titanium alloys

Pure titanium is generally used in implants where high strength is not necessary. For example, pure titanium is sometimes used to create fiber metal, a layer of metal fibers bonded to the surface of an implant that allows bone to grow into the implant or allows cement to better bond to the implant for stronger fixation.

Titanium alloys are biocompatible in nature. They commonly contain amounts of vanadium and aluminum in addition to titanium. The most used titanium alloy in knee implants is Ti6Al4V. Titanium and titanium alloys have great corrosion resistance, making them an inert biomaterial (will not change after being implanted in the body).

Titanium and its alloys have a lower density compared to other metals used in knee implants. Additionally, the elastic nature of titanium and titanium alloys is lower than that of the other metals used in knee implants. Because of this, the titanium implant acts more like the natural joint, and as a result, the risk of some complications like bone resorption and atrophy are reduced.

Uncemented implants

Knee implants may be cemented or cementless, depending on the type of fixation used to hold the implant in place. Most knee replacements are generally cemented into place. There are also implants designed to attach directly to the bone without the use of cement.

These cementless designs rely on bone growth into the surface of the implant for fixation. Most implant surfaces are textured or coated so that the new bone actually grows into the surface of the implant. For this, the surface of titanium is modified by coating the implant with hydroxyapatite, a bioactive surfacing agent that will ultimately bond as the bone grows into it.

Tantalum

Tantalum is a type of pure metal with excellent biological and physical properties, namely flexibility, corrosion resistance, and biocompatibility. Recently, a new porous substance has been made of tantalum and named *trabecular metal*. It contains pores, the size of which makes this material very good for bone in-growth. In addition, trabecular metal has an elastic nature that aids bone remodeling.

Polyethylene

The tibial and patellar components in knee replacements are made of polyethylene, which is a polymer. Though standard polyethylene surfaces traditionally have suffered from wear in hip implants, wear is less of a problem in knee implants, as the bearing surfaces are flatter and do not result in the same kind of wear. The use of ultra highly cross linked polyethylene (UHXLPE) or UHMWPE reduces even minimal wear, enabling the knee implants to last much longer.

Ultra-high molecular weight polyethylene

The most popular polymer used in orthopedics is UHMWPE. The principle stated by Charnley and Cupic [16] on low-frictional-torque arthroplasty consisting of either a hard metal or ceramic femoral head articulating against a UHMWPE acetabular cup with or without polymethyl methacrylate (PMMA) cement fixation is still followed in most hip replacements.

Zirconium alloy and all plastic tibial component

Zirconium alloy is used in a new ceramic knee implant. The zirconium alloy is combined with an all-plastic tibial component, replacing the metal tray and plastic insert used in other knee replacements. It is believed that this new knee could last for 20–25 years, substantially more than the 15–20 years that cobalt chromium alloy and polyethylene implants are effective. The new combination can be lubricated, which results in a smoother and easier articulation through plastic.

Another important characteristic of this material is that it is biocompatible, meaning that people with nickel allergies who cannot have knee implants made of cobalt chromium alloy (because nickel is an ingredient of cobalt chromium alloy) can have implants with this material. Zirconium alloy implants eliminate the risk to nickel-allergic patients because this new material contains no nickel.

Oxinium oxidized zirconium

Oxinium oxidized zirconium is a new material used in knee implants since 2001. It is basically a transformed metal alloy that has a ceramic bearing surface. It contains zirconium and niobium alloy that was oxidized to convert the surface of the material into zirconia ceramic. The advantage of this metal is that just the surface has been changed, so the rest of the implant component is a high tensile metal. Although it is twice as hard as cobalt chromium alloys, it provides half the friction, thus performing with higher quality and lasting for a longer time.

Ultimately, your knee replacement surgeon will recommend using whichever implant or implants he or she feels is right for your situation and whichever product he or she has previous success with. You should use this information to have informative preop conversations with your surgeon and to ask appropriate questions when investigating surgeons and surgery options.

1.6.1.2 Materials for hip arthroplasty

Hip replacement is based on a ball and socket joint. The femoral stem and ball fit into and have relative movement against the cup or acetabular component, as shown in Fig. 1.6. There are a variety of materials from which each of the components can be fabricated. Each manufacturer has different models, but each style falls into one of four basic material categories (refer to https://bonesmart.org/):

- metal on plastic (polyethylene or UHMWPE)
- metal on metal (MoM)
- ceramic on plastic (UHMWPE)
- ceramic on ceramic (CoC)

Metal on plastic

This combination has been in use in various forms since some of the earliest hip replacements back in 1960 (when it was called the low friction arthroplasty (LFA)). Some years later, the make-up of the polyethylene was improved. The current plastic used in hip replacement implants is referred to as ultra highly cross-linked polyethylene

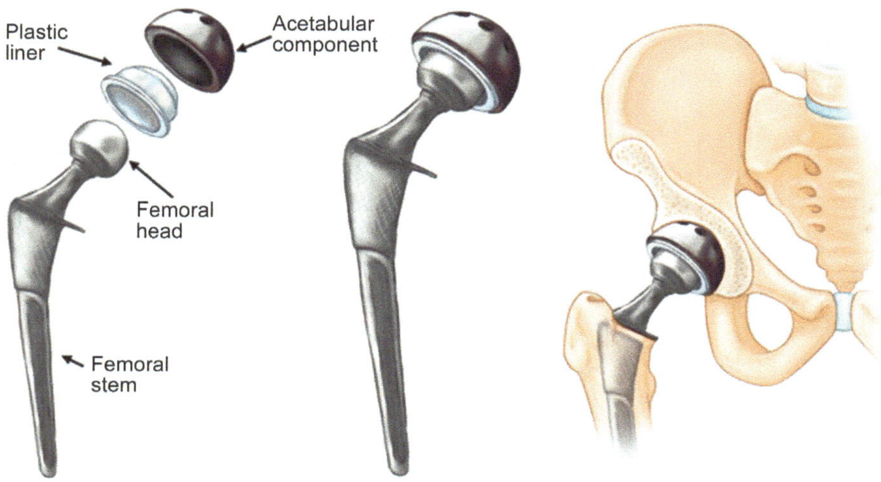

Fig. 1.6 (*Left*) Individual components of a total hip replacement; (*Center*) Components merged into an implant; (*Right*) Implant as it fits into the hip.
From http://orthoinfo.aaos.org/topic.cfm?topic=a00377. Reproduced with permission from OrthoInfo. © American Academy of Orthopaedic Surgeons. http://orthoinfo.aaos.org.

(UHXLPE) or ultra high molecular weight polyethylene (UHMWPE), a very stable and reliable plastic material with greatly reduced risk for wear.

Because of its durability and performance, metal on polyethylene has been the leading artificial hip component material chosen by surgeons since hip replacement surgeries were first performed. It is also the least expensive bearing.

All implants shed debris as they wear. Over time, the body may see polyethylene wear particles as invaders or a source of infection. As the body starts to attack them, this leads to osteolysis, a "dissolving of the bone," which may result in having to replace the implant (known as revision).

As noted previously, technological advances have reduced the risk of wear in metal-on-polyethylene implants. They wear at a rate of about 0.1 mm each year. The other materials, metal and ceramic, being more modern developments, already have high wear resistance built in.

Metal on metal

Metal-on-metal (MoM) hip implants have been used even longer than metal-on-plastic implants. MoM bearings (cobalt chromium alloy, titanium alloy, or sometimes stainless steel) were in use from as far back as 1955, though they were not approved for use in the United States by the FDA until 1999. They offer the potential for greatly reduced wear, with less inflammation and less bone loss. Some device recalls have brought negative attention to MoM.

Metal bearings are available in many sizes (28–60 mm); there are also several neck lengths available. Only metal-on-metal components allow the largest heads throughout the entire range of implant sizes. Large ball heads provide increased range of motion and greater stability, which can significantly reduce the risk of hip dislocation, a crucial factor in the long-term success of an implant.

Because the human femoral (ball) head is naturally large, it makes sense to implant a large, anatomic replacement. This was not possible in the past because traditional design parameters made smaller femoral heads necessary. However, with the introduction of metal-on-metal implant components, liners may be eliminated, allowing surgeons to use large femoral heads.

MoM implants have a potential wear rate of about 0.01 mm each year. Although wear is reduced with MoM implants, the wear products (submicroscopic particulates, soluble metal ions) are distributed throughout the body. This has raised concerns about long-term biocompatibility. At present these are only concerns, for there have been no definitive clinical findings that these wear products are harmful. It should also be noted that this issue arises fairly rarely.

Ceramic

Ceramic is the 21st century's answer to hip replacement, as it is both hard and durable, it wears minimally, and the material is widely deemed to have no toxic or side effects in the human body. Hip implants can be constructed as ceramic-on-UHMPE or ceramic-on-ceramic (CoC).

Ceramic on ceramic

If you are a very active individual or a relatively young patient, your surgeon may prescribe an all-ceramic hip joint. CoC is a good combination with longevity and reliability. In these hip joints, the traditional metal ball and polyethylene liner are replaced by a high-strength ceramic bearing that has a reputation for ultra-low wear performance. Clinical studies, monitored by the FDA and begun in 1998, have demonstrated excellent performance, although it should be noted that ceramic was used in hip replacements for many years prior to that.

All-ceramic hip joints have been used in Europe since the 1980s but have only more recently received the FDA's approval for marketing in the United States. There was a history of two issues with ceramic hips: catastrophic shattering and squeaking. Shattering was more of an issue in the 1980s and 1990s but the product has been substantially improved since then, essentially eradicating the shattering problem. Squeaking, however, remains a bit of a problem for a few patients. Often the noises abate over time but sometimes they don't. If the squeaking is intolerable, a revision may be necessary.

Ceramic is the hardest implant material used in the body, and has the lowest wear rate of all, to almost immeasurable amounts (1000 times less than metal-on-polyethylene, about 0.0001 mm each year). Consequently, there is usually no inflammation or bone loss, nor systemic distribution of wear products in the body. New ceramics offer improved strength and more versatile sizing options.

Ceramic on plastic (or UHMWPE)

Ceramic on UHMWPE is a good combination of two very reliable materials. Ceramic heads are harder than metal and are the most scratch-resistant implant material. The hard, ultra-smooth surface can greatly reduce the wear rate on the polyethylene bearing. The potential wear rate for this type of implant is less than for metal on polyethylene.

Ceramic on polyethylene is more expensive than metal on polyethylene, but less than CoC. In the past, there were incidents of fractures in ceramic components, but newer, stronger ceramics have resulted in considerable reduction of fracture rates (0.01%) compared to the original, more brittle ceramics.

Some ceramic-on-polyethylene implants utilize a vitamin E-stabilized, highly cross-linked polyethylene bearing material. Vitamin E, a natural antioxidant, is expected to improve the longevity of the implant bearings used in total joint replacements. In laboratory testing, these liners have demonstrated 95%–99% less wear than some other highly cross-linked polyethylene liners.

Ceramic-on-polyethylene implants have a potential wear at a rate of about 0.05 mm each year, i.e., 50% less than metal on polyethylene. The newer, highly cross-linked polyethylene liners have shown potential wear rates as little as 0.01 mm each year.

1.6.2 Emphasis on titanium alloys for making implants

Titanium (Ti) alloys are metallic materials that contain a mixture of Ti and other chemical elements. These alloys are known for their low density, high specific strength, high melting temperature, and superior corrosion resistance [17]. However, issues like high friction coefficient, high sensitivity to adhesive wear and fretting wear, as well as their poor resistance to high temperature oxidation and low wear resistance, have restricted the application of Ti alloys. Many surface modification techniques have been developed to increase the usability of these alloys. Some of the notable techniques include laser surface modification technology, anodic oxidation, ion implantation, and deposition.

Titanium alloys can be applied as biomedical materials mainly in hard tissue replacement, which requires a modulus similar to human bone and high fatigue strength. The typical mechanical properties of biomedical Ti-based alloys are given in Table 1.2, while the fatigue strengths at 10^7 cycles are shown in Fig. 1.7. It is observed that the fatigue strengths displayed by Ti-based alloys can reach as high as 800 MPa (for Ti-6Al-4V ELL).

Table 1.2 Mechanical properties of Ti alloys for biomedical applications [17]

Ti alloy	Ultimate tensile strength (MPa)	Yield strength (MPa)	Elastic modulus (GPa)	% Elongation
Pure Ti	240–550	170–485	102.7–104.1	24–15
Ti-6Al-4V ELI (mill annealed)	860–965	795–875	101–110	6–10
Ti-6Al-7Nb	900–1050	880–950	114	8.1–15
Ti-5Al-2.5Fe	1020	895	112	15
Ti-5Al-1.5B	925–1080	820–930	110	15–17
Ti-13Nb-13Zr (aged)	973–1037	836–908	79–84	10–16
Ti-12Mo-6Zr-2Fe	1060–1100	100–1060	74–85	18–22
Ti-15Mo-5Zr-3Al	852–1100	838–1060	80	18–25
Ti-15Mo-2.8Nb-0.2Si	979–999	945–987	83	16–18
Ti-29Nb-13Ta-4.6Zr	911	864	80	13.2

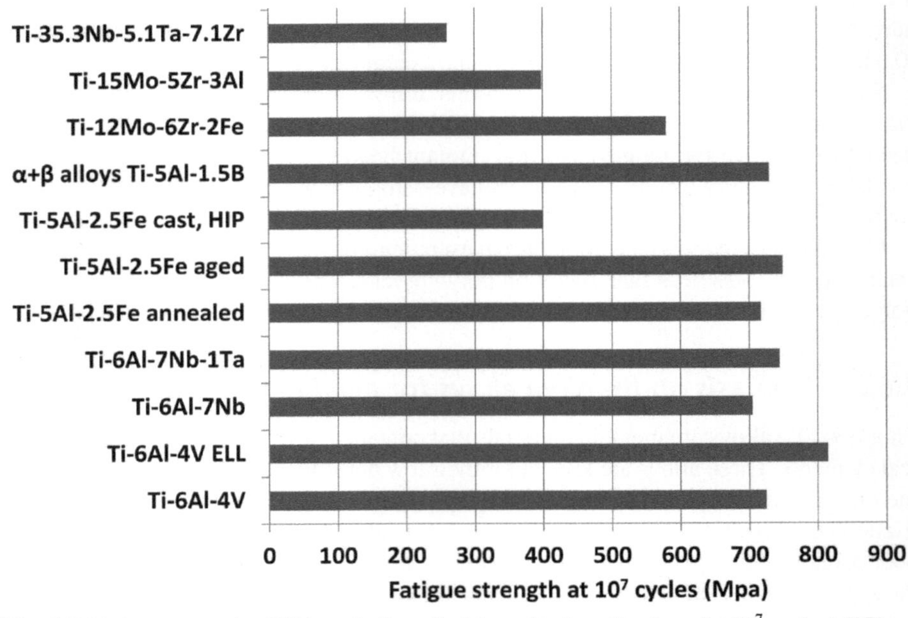

Fig. 1.7 Fatigue strength of Ti-based alloys for biomedical applications (at 10^7 cycles) [17].

1.6.2.1 Biocompatibility of titanium alloys

Biocompatibility is the ability of a material to remain biologically innocuous to living tissue or when inserted inside a living body. According to IUPAC, biocompatibility is the ability of a material to be in contact with a living system without producing an adverse effect. Biocompatible materials don't produce a toxic or immunological response when exposed to the body or body fluids. Moreover, the materials should not biodegrade within its useful life as an implant or prosthetic. Any interaction between the implant and the living body may lead to injury, which is generally of the following types:

(i) Release of metallic ions from the implant, which may interfere with physiological movement of ions in neural cells.
(ii) Concentration of metallic ions in the human body that may be the cause of many adverse local tissue reactions. The corrosion rate of the implanted alloy and the solubility of the corrosion products determine the release of the ions.

Hence, monitoring of released corrosion products and their stability in tissues is very significant. The alloy elements, namely tantalum, niobium, vanadium, and zirconium, produce essentially insoluble oxides and their corrosion products can have good stability in the human body.

1.6.2.2 Osseointegration in titanium alloys

As part of biocompatibility, materials intended for biomedical applications should also possess excellent osseointegration. Osseointegration is the direct structural and functional connection between the surface of the implant and the living bone in the human body. As is evident, osseointegration requires a high level of biocompatibility so that the foreign material becomes a part of the living system. In the case of Ti-based alloys, several in vivo and in vitro studies have been carried out. Nails made of Ti (Ti6Al4V) are implanted into healthy and osteopenic rodents and in vitro studies also used cells cultured from the same animals [18]. After a couple of months, it was found that the osseointegration of Ti6Al4V happened in normal and osteopenic bone, and the rate of osseointegration in normal bone is greater than that in osteopenic bone. The effect of surface roughness of the implant on the bone remodeling activity was studied and it was found that a rough surface supplies a stable bone-implant interface and promotes the osseointegration of Ti implants [19]. A special surface treatment method has been used to promote bone growth on the Ti surface by increasing the bone-implant interface. More recently, some elements, such as Nb, Ta, Zr, and Mo, have been used as alloying elements to develop some new alloys, such as Ti22Nb6Zr. Those new alloys present a good electrochemical behavior in physiological fluids and can induce better osseointegration; as a result, they are expected to become promising candidates for biomedical applications.

1.7 Tribological testing of biomaterials

As biomedical materials are to be used directly within a living body, their thorough testing is required so that they perform satisfactorily over the entire duration of their life. Tribological testing of artificial joints is carried out extensively in laboratories under simulated conditions with various degrees of complexity before the implants are approved for clinical applications. For consistency of the tests, a number of standards for wear testing of artificial joints have been introduced, viz. ISO (14242, 14243, and 18192) and ASTM (F2025). A major challenge in testing of these components is to create simulating conditions very close to the actual clinical scenarios. This includes subjecting the joint to more adverse conditions to reflect a wide spectrum of use in patients and by surgeons [20]. Conditions such as changes in loading and activity, third-body wear, surface topography, edge wear, and the role of aging of the bearing materials are taken into account.

1.7.1 Tribometry configurations

Tribometry, in general, represents an area of tribology that encompasses means and methods of measuring: friction forces in contact zones; wear of tribosystem elements; temperature; surface roughness; contact surface sizes; contact strain etc. [21]. Tribometers are devices used to evaluate a material's tribological properties, including friction, wear, and even adhesion, hardness, and other contact parameters. Choubey et al. [22] have employed a ball-on-flat fretting wear tester (Fig. 1.8) to evaluate the tribological behavior of various titanium alloys. Fretting is defined as low-amplitude reciprocating tangential sliding. It is noted that a majority of the tribological tests involving biomaterial combinations are subjected to standard reciprocating motion similar to real contact conditions prevalent in the human body [22]. The wear of the sample is measured either through the conventional weight loss technique or directly from the instrument using a displacement transducer. The friction force is recorded with the help of a load cell and the coefficient of friction is obtained by dividing the friction force by the normal force exerted by the counter body. Besides this, common tribotesting configurations (Fig. 1.9), namely ball-on-disc, pin-on-disc, block-on-disc, etc., are employed to evaluate the tribological characteristics of biomedical materials, including metals and alloys, UHMWPE, ceramics, etc. Either the ambient temperature or the human body temperature (about 37°C) is selected as the working temperature of the tests. Moreover, tests are carried out generally under lubrication, where the lubricant used is simulated body fluids. Some tests are performed in a dry condition also.

Thanks to advancements and developments in the field of medicine and surgery, approximately one million total hip replacement surgeries are performed worldwide annually [23]. The lifetime of these implants could be as high as up to 30 years for elderly patients who remain mostly inactive. On the other hand, in younger and more active patients, the prosthesis will be subjected to higher stress and wear. Thus there is a continuing and urgent need to test newer materials and joint designs, so as to increase the wear resistance of such joint implants. Apart from the regular tribological tests, some special tribological experiments have been carried out simulating the actual implant conditions. Hip and knee wear machines are used to simulate the performance of these materials under conditions believed representative of the patient's gait and environment (Table 1.3). However, one of the difficulties in evaluating various simulator wear studies is the lack of consistent test parameters. Simulators can be single-axis,

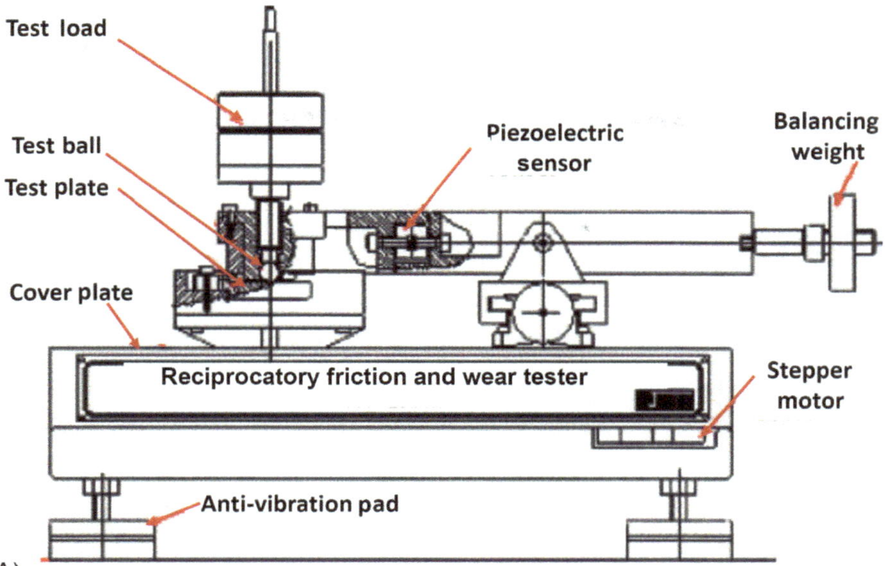

(A)

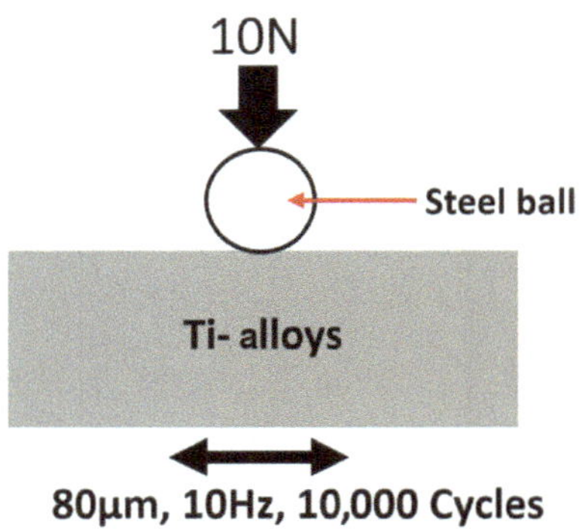

(B)

Fig. 1.8 (A) Schematic of the fretting wear tester, and (B) schematic representation of the fretting of Ti-based alloy against steel.
(A) Source: DUCOM, India. (B) Source: A. Choubey, B. Basu, R. Balasubramaniam, Tribological behavior of Ti-based alloys in simulated body fluid solution at fretting contacts, Mater. Sci. Eng. A 379 (1–2) (2004) 234–239.

orbital 2-axis, and 3-axis machines. In addition, different types of lubricants are used in the tests.

Brown and Clarke [23] have reviewed the results of a multiaxial hip simulator set-up (shown in Fig. 1.10), which can easily replicate the motion of the human hip. There are three types of motion in hip simulator designs:

(a) Produces only the flexion-extension motion of the hip joint [24].
(b) Orbital machines employ a biaxial rocking motion. They have a standard (±23 degree) orbiting cam that provides a synthesis of the three-dimensional motions of the human hip joint.

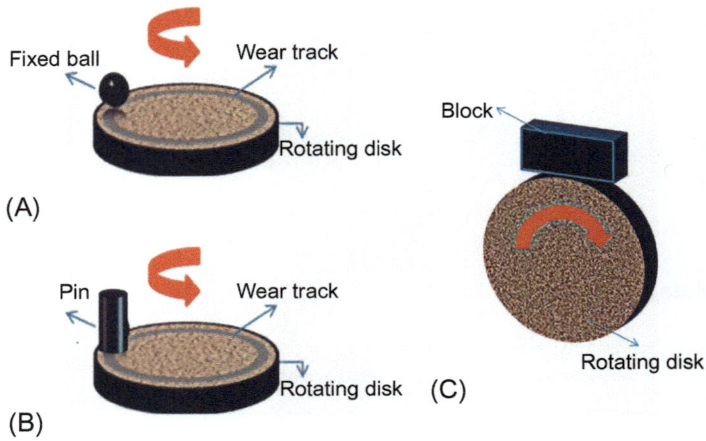

Fig. 1.9 Schematic of (A) ball-on-disc, (B) pin-on-disc, and (C) block-on-disc wear test configurations [3].

Table 1.3 Operating parameters in the human hip compared to the hip simulator [23]

Sl. no.	Parameter	Human hip (actual)	Human hip (artificial)
1	Joint	Ball and socket (40–60 mm)	Ball and socket (22–32 mm)
2	Velocity	0–127 mm/s	0–127 mm/s
3	Flexion	46 degrees	46 degrees
4	Motion	Three dimensions (intermittent, variable)	Three dimensions (runs nonstop)
5	Loading	0–2 kN	0–2 kN
6	Bearing	Cartilage on cartilage	Metal on polyethylene
7	Lubricant	Synovial fluid	Water, saline, serum, other
8	Temperature at interface	0–2°C change	12–70°C change possible
9	Volume	2 mL	200–500 mL
10	Viscosity	Medium to high (non-Newtonian)	Low to medium (Newtonian)
11	Replenishment	Continuous	At intervals
12	Degradation	Not known	Extensive

(c) Triple-axis machines have provision for flexion, internal/external rotation, plus abduction and adduction motions [25]. A metal ball oscillates against a hemispherical cup of UHMWPE to mimic the human hip.

The position of the acetabular cup can be fixed in several ways relative to the machine axes. Either the ball or the cup can be fixed and the other remains free to rotate. The cup may be fixed above the femoral ball (anatomical), as it is in the patient, or it may be mounted below the femoral ball (inverted) to ensure more consistent wetting of the bearing surfaces [23]. Generally, the hip joint is surrounded by a plexiglass chamber or flex sleeve, which holds the fluid lubricant. A physiological load profile

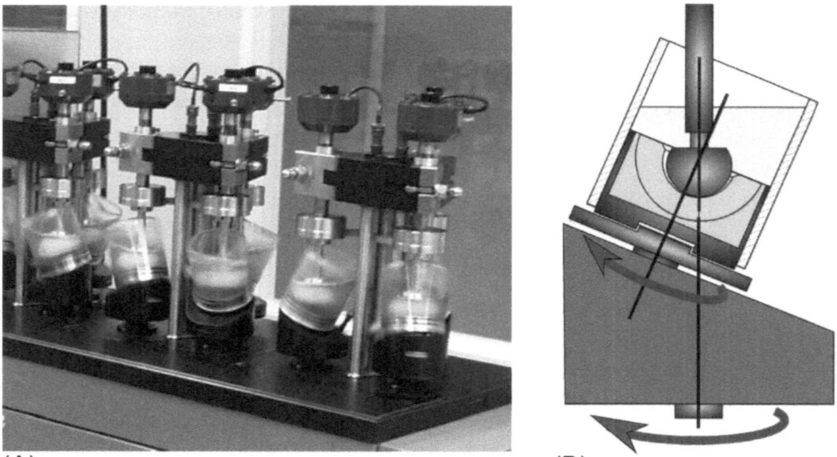

Fig. 1.10 (A) Biaxial orbital hip simulator set-up; (B) Schematic of simulator set-up [23].

mimics the in vivo walking load [26] most commonly using only one axis for the resultant hip force. With a frequency of typically 1 Hz, most studies of UHMWPE cups had a two to five million cycle duration. In actual cases, patients have averaged one million cycles per year [27].

1.7.2 Tribometry at small scales

It is critical to measure and calculate the friction force of biological implants for human joints, where friction force defines the system behavior and is also responsible for wear. However, there are some differences between the tribo testing of biomaterials compared to tests conducted in metal-working industries. This is because biomaterials are mostly softer than bare metals and alloys. Moreover, investigations on biomaterials should be multidisciplinary, taking into account diversified data. Anticipating the wear mechanism that the components would be subjected to and accordingly devise the testing model remains one of the major challenges in testing. Accurate surface modeling, using various aspects of contact type, temperature, lubricating modes, and environment, is also crucial for valid testing [21]. It is difficult to set up simulation models for laboratory testing of tribological systems in relation to a human body. Investigations are currently being carried out to establish more accurate methodologies for predicting wear in complex environments of existing biotribological systems. This would greatly enhance the development of more durable biomaterials for application in human systems. Many of the most important advances of recent years have come from new techniques capable of characterization at small scales, even at the level of individual molecules. In the case of biotribology, friction estimation on micro- and nanoscales, experimental results for controlled load, and wear tests of polymer film can all be realized using nanotribometer instruments, thus contributing to the previously mentioned issues. In these cases, laboratory simulations as well as in situ testing are needed. Laboratory tests are necessary to help optimize biomaterial performance, among which nanotribometer testing can be very significant for extended wear testing. Many researchers have investigated wear debris and its effect on the system in question, leading to identification of ample numbers of material pairs (metal on metal, ceramic on metal, etc.), as well as improvement of existing contact systems by application of surface engineering technologies (e.g., multilayered coatings). However, the need still exists to study these systems from the point of view of their long-term clinical performance.

1.7.3 Lubrication under body fluids

Lubrication and wear mechanisms acting within biotribological systems (e.g., joints) are still not fully understood and need to be investigated further. Usually the components are subjected to boundary lubrication in human body fluids. Synovial fluid is the chief biological joint lubricant. However, it is present in quantities too small to be used for wear simulation studies. Hence, a variety of liquids are used, starting from plain water and moving to saline solutions, synovial fluid components (hyaluronic acid, lubricin, and phospholipids), serum, and custom-made lubricants [23]. In the case of metal on polyethylene testing, serum is used extensively as the lubricant because its results closely match clinical results with polyethylene bearings.

1.7.3.1 Water as lubricant

Initial simulator studies employed distilled water or a saline solution as the lubrication fluid. The reason behind this was the easy availability and the fact that they are inexpensive and are not prone to degradation and bacterial contamination. However, simulator studies using water show wear rates inconsistent with clinical hip wear, polyethylene transfer, and debris size and shape. In the presence of water, transfer of polyethylene from the liners to the balls is observed. Wear particles generated in experiments with water as lubricant are massive flake-like particles, some of which are several millimeters across. However, wear debris retrieved from periprosthetic tissues removed during revision of total hip replacements indicated that the vast majority of the wear particles were submicron sized and were either rounded or elongated in shape [23]. Thus, tests with water as lubricant are no longer used.

1.7.3.2 Saline solution as lubricant

Ringer's solution is a saline solution with the same advantages as water. However, it has salt content quite similar to that of natural lubricant [28]. Some researchers have reported that saline wear results were almost the same as those in water [29,30]. Similarly to water, polyethylene transfer is observed with the use of saline solution. However, the transferred layer of polyethylene had a definite orange cast, suggesting some corrosion, and the transferred film would occasionally break up into flakes of debris [31].

1.7.3.3 Bovine serum as an implant lubricant

Bovine serum has a resemblance to the body fluid of human beings and hence is currently the most commonly used lubricant in joint simulator machines [32]. Wear under bovine serum is of the same order of magnitude as clinical wear. There is visibly no polyethylene transfer when serum is the lubricant and testing under serum lubrication produced extremely small wear particles, mostly submicron sized and either rounded or elongated as in retrieved debris [32]. Bovine serum cannot be treated as the perfect lubricant, as it is expensive and degrades fast. Its viscosity was lower than synovial fluid and the viscosity did not change with the shear rate, making it a Newtonian fluid. Moreover, the lubricating properties of serum are complex and not yet fully understood.

1.7.3.4 Bovine synovial fluid as an implant lubricant

Bovine synovial fluid (BSF) is not commercially available. For using BSF, the synovial fluid just after slaughter is collected. Care should be taken the fluid is not visually contaminated with blood. After collection the fluid is properly centrifuged to remove the cellular debris and then subjected to ultrafiltration. The ultrafiltration retentate is then fractionated. As the total process takes almost a week, BSF is not a practical lubricant for implants.

1.7.3.5 Pseudo-bovine lubricant

Many substitutes for the expensive bovine serum have been experimented with. A gelatin-based solution called Gelofusine (also known as modified fluid gelatin) remained uncontaminated by bacteria in the simulator. In addition, the wear rate was quite similar to that in bovine serum. However, the wear debris was much larger and particle numbers lower, very similar to studies in water. Another modified fluid gelatin called Plasmion has been used in France [23]. It is a synthetic serum with a protein content of 30 mg/mL. Unfortunately, wear results were similar to those using water as the lubricant. Carboxy methyl cellulose (CMC) used in studies of implant friction showed a strong correlation between experimental results and theoretical predictions of film thicknesses and lubrication modes. Thus it seems that a clinically relevant synthetic lubricant has yet to be discovered.

1.8 Tribological properties of materials for biomedical applications

1.8.1 Metallic biomedical materials

1.8.1.1 Stainless steel (first generation)

Stainless steel (SS) was one of the first-generation metallic materials to be used for biomedical applications [33]. The main benefit of SS was its corrosion-resistant properties in various environments. While many types of SS are available for making implants, 316L (ASTM F138, F139) has been the most popular one. The carbon percentage in this steel is <0.03 (wt.%) with chromium (17%–19%) and nickel (12%–14%) being the major alloying elements in iron. The presence of chromium results in the development of a corrosion-resistant oxide film of Cr_2O_3, which adheres strongly to the surface of steel. Moreover, the low carbon content decreases the chances of the formation of carbides, viz. $Cr_{23}C_6$, which tend to precipitate at grain boundaries under favorable conditions. This formation disturbs the grain boundaries of chromium and hence makes the steel prone to corrosive attack.

With long-term use of 316L-based implants in the human body, corrosion is found to occur, which leads to the release of Fe, Cr, and Ni ions, which are possibly powerful allergens and carcinogens. Studies show that more than 90% of the failures of implants of 316L stainless steel are due to pitting and crevice corrosion attacks [3]. Fig. 1.11 shows the corrosion occurring on the surface of a SS implant [34].

To prevent allergic reactions, research is also underway to develop Ni-free SS. Austenitic steel, namely 316L, also has poor tribological properties, especially in situations where wear can be responsible for material degradation. The tribological properties of the steel could be enhanced by surface treatment. However, with the advancement of material science and development of newer metals, SS has almost been obliterated from the field of biomedical engineering.

1.8.1.2 Cobalt chromium alloy (second generation)

The cobalt chromium alloys are the second-generation implant metals, which are widely used in artificial knee and hip joints due to their wear resistance characteristics [33]. The wear resistance of cobalt-based alloys is even higher than SS and even titanium-based alloys [35]. High strength cobalt alloys, such as Co-Cr-Mo alloys, are used to fabricate hip joints which are subjected to wear. It has been reported by Niinomi [36] that diffusion of carbide in cobalt alloys increases the wear resistance of the alloy. Although the carbide formed significant features, these features did not affect the corrosion performance of the alloy. The transformation of the metastable gamma phase to the martensitic phase (via a deformation-induced transformation) has been found to improve the wear resistance of cobalt alloys.

There are three different types of Co-Cr-Mo material currently in use: cast (low carbon), wrought, and wrought (high carbon) alloys. Each variety has a different

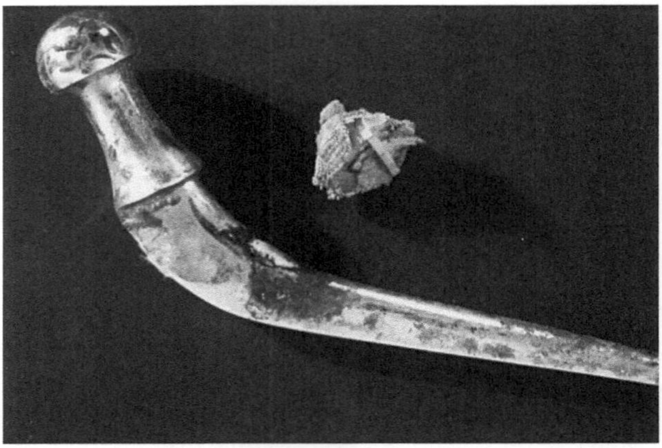

Fig. 1.11 Failure by corrosion for SS implant [34].
From J. Walczak, F. Shahgaldi, F. Heatley, In vivo corrosion of 316L stainless-steel hip implants: morphology and elemental compositions of corrosion products, Biomaterials 19 (1998) 229–237.

microstructure and different properties optimized for a specific design or function [4]. Wrought Co-Cr alloys are found to possess higher strength compared to cast Co-Cr alloys and hence the former alloys are employed in making implant devices with high strength requirements. The nickel content in Co-Cr alloys triggers allergic reactions. Hence, nickel-free alloys are developed for implant-based applications.

Corrosion studies under body fluid showed that cobalt dissolved from the surface of the alloy and the remaining surface oxide consisted of chromium oxide (Cr^{+3}) containing molybdenum oxide (Mo^{+4}, Mo^{+5}, and Mo^{+6}) [37]. It is reported that increase in the level of cobalt in the blood is found in the patient during the first year of the implant [4]. However, the origin of this free molybdenum may be due to wear or corrosion. Several surface treatment techniques, e.g., surface coating, ion implantation, and doping, have been evolved to enhance the wear resistance, reduce the friction coefficient, and increase the hardness of Co-Cr based alloys.

1.8.1.3 Titanium and titanium-based alloys (third generation)

Titanium is gaining as one of the most promising arthroplasty implant materials due to its excellent mechanical and tribological properties. Titanium and its alloys possess superior biocompatibility and excellent corrosion resistance because of the thin oxide layer formed on the surface. They are used commonly in artificial knees and hip joints. The presence of this oxide film that forms spontaneously in the passivation or repassivation process is a major criterion for the excellent biocompatibility and corrosion resistance of titanium and its alloys [38].

The use of pure titanium is more limited to dental implants due to its limited mechanical properties. In knee and hip implants, bone screws, and plates, where enhanced mechanical properties are desired, an alloy of titanium (Ti-6Al-4V) is used [39]. One of the most common applications of titanium alloys is artificial hip joints that consist of an articulating bearing (femoral head and cup) and stem (shown in Fig. 1.12), where metallic cup and hip stem components are made of titanium alloy. Titanium alloys are also often used in knee joint replacements as well, which consist of a femoral and tibial component made of titanium and a polyethylene articulating surface (Fig. 1.12).

Fig. 1.13 plots the steady-state coefficient of friction (COF) as a function of different material combinations. The standard deviation of the COF for at least three fretting experiments is represented by the error bars. Insignificant variation in the friction values (0.01–0.02) is noted for all the materials. It is found that commercially pure (CP) titanium displays the highest COF (around 0.5) while the lowest COF is noticed for Ti-5Al-2.5Fe (around 0.30) among all the alloys. The COF values for Ti-13Nb-13Zr and Ti-6Al-4V are comparable at 0.48 and 0.46, respectively. Co-28Cr-6Mo (steel) exhibits a steady-state COF of around 0.4.

Although titanium possesses a host of advantages, the surfaces of the metal and its alloys have relatively poor wear resistance. In particular, titanium surfaces in contact with each other or with other metals readily gall under conditions of sliding contact or fretting. Even with light loading and little relative movement, complete seizure of the surfaces can occur. This situation is caused by adhesive wear in which microscopic asperities on the metal surfaces come into contact due to relative sliding, and they tend to weld together, forming a bond at the junction that can have rupture strength

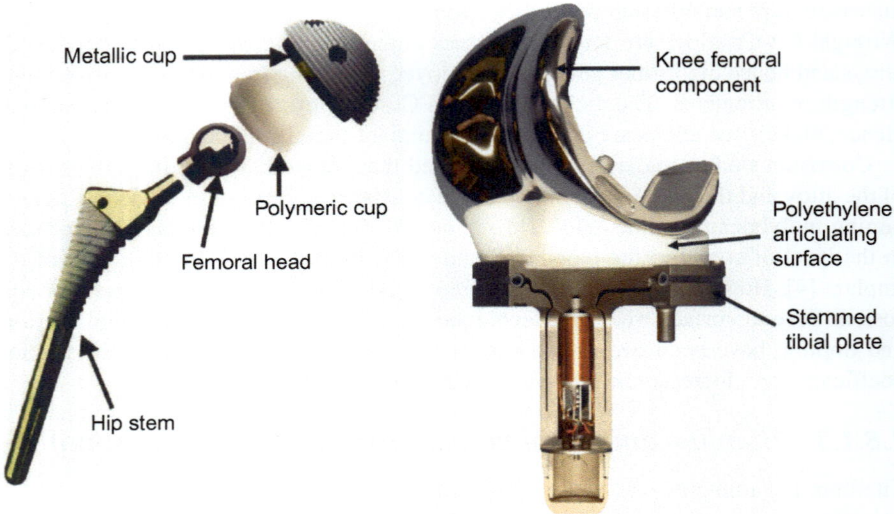

Fig. 1.12 Schematic diagram of artificial hip joint (*left*) and knee joint (*right*). (*Left*) From V.S. Viteri, E. Fuentes, Titanium and titanium alloys as biomaterials, in: J. Gegner (Ed.), Tribology—Fundamentals and Advancements, InTech, Rijeka, 2013. (pp. Ch. 05). (*Right*) From https://upload.wikimedia.org/wikipedia/en/9/99/Knieprothese.png.

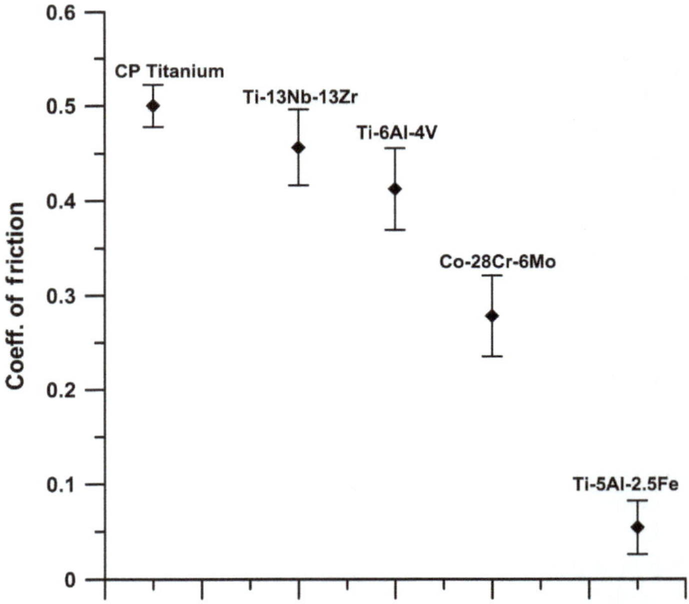

Fig. 1.13 Variation of steady-state coefficient of friction for different materials measured during fretting against 8 mm diameter steel ball at 10 N load for 10,000 cycles with a frequency of 10 Hz for 80 m displacement stroke [22].

greater than the strength of the underlying metal. Fracture then takes place at one of the asperities, causing metal to be transferred from one surface to the other. The debris so formed gives rise to the accelerated wear that occurs with titanium. Due to this, in the case of total joint replacements made with a titanium head and polymer cap, 10%–20% of joints need to be replaced within 15–20 years and the aseptic loosening accounts for ~80% of the revised surgeries [38]. The wear debris released from the implant gets into the bloodstream, resulting in inflammation of the surrounding tissue and giving rise to osteolysis (pathological destruction of bone tissue). This finally leads to implant loosening, which then needs replacement.

One of the important criteria that any biomedical material should possess is corrosion resistance, as the implants are in continuous contact with body fluids at different pH levels. It has been documented in the literature that the average pH value of the fluids in the human body is about 7.4, which may increase (up to 7.8) after a surgery and then return to its normal value after a few days [17]. Under this environment, Ti and its alloys exhibit higher corrosion resistance compared to cobalt-based alloys and stainless steel (316L). This corrosion resistance property of Ti alloys can be attributed to a stable passive layer formed due to reactions between Ti alloys and body liquids consisting of water molecules, dissolved ions, and proteins. X-ray diffraction of the layers indicated the formation of orthorhombic TiO_2 films on the surfaces of pure Ti as well as some of the Ti alloys (Ti6Al4V and Ti-10%Ta) [40].

Several studies have highlighted the variation in the corrosion performance of Nitinol depending upon the surface condition of the test specimens used and the surface condition given. Since heat treatment is involved during the manufacturing process, the passivating oxide present on Nitinol is polycrystalline in nature and has been found to exhibit severe pitting and crevice corrosion, whereas surface treatment to form amorphous oxide results in excellent corrosion resistance.

Common mechanical properties of some of the metallic biomaterials are compared with bone in Table 1.4. It is found that the elastic modulus of Co-Cr alloys and stainless steel is 10 times that of the bone, which may cause stress shielding. However, the Young's modulus of titanium and its alloys is ~0.5 times that of stainless steel, and hence the risk of stress shielding is less in titanium and its alloys compared to Co-Cr alloys and stainless steel. Examples of the metallic alloys used in biomedical applications, along with their advantages and disadvantages, are summarized in Table 1.5. Different wear studies for metallic biomedical materials as compiled by Hussein et al. [3] are presented in Table 1.6.

Table 1.4 Comparison of mechanical properties of popular metallic biomaterials with bone [41]

Material	Young's Modulus (GPa)	Yield Strength (MPa)	Tensile Strength (MPa)	Fatigue Limit (MPa)
Stainless steel	190	221–1213	586–1351	241–820
Co-Cr alloys	210–253	448–1606	655–1896	207–950
Titanium (Ti)	110	485	760	300
Ti-6Al-4V	116	896–1034	965–1103	620
Cortical bone	15–30	30–70	70–150	–

Table 1.5 Comparison of metallic biomaterials applied in human body [3]

Metals and alloys	Selected examples	Advantages	Disadvantages	Principal applications
Titanium-based alloys	CP-Ti Ti-Al-V Ti-Al-Nb Ti-13Nb-13Zr Ti-Mo-Zr-Fe	Low Young's modulus excellent, corrosion resistance, low density, high biocompatibility	Poor tribological properties, toxic effect of aluminum and vanadium in long term	Bone and joint replacement, fracture fixation, dental implants, pacemaker encapsulation
Cobalt and Cr alloys	Co-Cr-Mo Cr-Ni-Cr-Mo	High wear resistance	Higher modulus than bone, allergy possibilities with nickel, chromium and cobalt	Bone and joint replacement, dental implants, dental restorations, heart valves
Stainless steels	316L	High wear resistance	Higher modulus than bone, allergy possibilities with nickel, chromium and cobalt	Fracture fixation, stents, surgical instruments
Others	Ni-Ti	Low Young's modulus	Allergy to nickel	Bone plates, stents, orthodontic wires
	Platinum and Pt-Ir	High corrosion resistance under extreme voltage potential and charge transfer condition		Electrodes
	Hg-Ag-Sn amalgam	Easy in situ formability to a desired shape susceptible to corrosion in the oral environment	Toxicity due to mercury	Dental restorations

1.8.2 Ceramic biomedical materials (bioceramic)

Ceramics are commonly implemented as dental and bone implants due to biocompatibility and wear resistance. They can also act as bone replacements and can be used as supports for the regeneration of tissues due to their porosity [33]. The class of ceramics used for repair and replacement of diseased and damaged parts of musculoskeletal systems are termed bioceramics. Other medical uses of bioceramics are in pacemakers, kidney dialysis machines, and respirators. There are three classes of bioceramics, as follows:

(a) Bioinert: Completely inert towards the living tissue with no interaction, e.g., alumina (Al_2O_3), zirconia (ZrO_2), SiC, Si_3N_4, etc.

Table 1.6 Compilation of different wear studies on metallic biomaterials [3]

Material and fabrication processes	Experimental test techniques and parameters	Main results	Study/ reference
Ti-13Nb-13Zr Ti-6Al-4V Arc melting	A **block-on-disc** tribometer was used to conduct wear and friction tests in a simulated body fluid (Ringer's solution) **Temperature**: ambient **Normal load**: 20–60 N **Sliding speed**: 0.26–1.0 m/s	The Ti-6Al-4V alloy showed a higher wear resistance than the Ti-13Nb-13Zr alloy. Abrasion was the primary wear mechanism	[42]
NiTi commercial alloy	A **block-on-disc** was used to measure dry sliding wear. A profilometer was used to quantify wear **Sliding speed**: 0.837 m/s **Sliding distance**: 1004 m **Loads**: 50–200 N	A NiTi/WC-Co coupling exhibited a high wear rate. A transition from delamination wear to a regime featuring a mixture of delamination and oxidation wear	[43]
Ti-13Nb-13Zr equal channel angular pressing (ECAE)	A tribometer was used as a lubricity fretting test system for texture and wear behavior; fretting wear and 3D surface texture measurements were performed **Normal loads**: 6 N **Frequency**: 20 Hz **Temperature**: 37 ± 0.1°C	The grain size and texture of material affected the wear of the surface. There was no difference in the friction coefficient between the ECAE processed and as-received samples	[44]
Ti-6Al-7Nb and AISI 316L stainless steel	**Ball-on–disc** and sphere-on-plane **Load**: 3, 6, and 10 N **Sliding speed**: 1, 15, and 25 mm/s	The same mechanisms of wear and friction were found for all tested samples	[45]
Ti-Nb-Ta-Zr and Ti-6Al-4V induction skull melting method	**Reciprocal pin-on-disc** in a 0.9% NaCl solution **Reciprocating velocity**: 45 rpm **Sliding distance**: 30 km	The wear resistance of Ti-29Nb-13Ta-4.6Zr was enhanced by incorporating Nb2O5 oxide particles into the diffusion-hardened surface of the alloy	[46]
CP titanium, Ti-6Al-4V, Ti-5Al-2.5Fe, Ti-13Nb-13Zr and Co-28Cr-6Mo	**Ball on flat fretting** wear tester: Hanks' balanced salt solution **Normal load**: 10 N for 10,000 cycles **Frequency**: 10 Hz	The primary wear mechanisms of Ti alloys were tribomechanical abrasion, transfer layer formation, and cracking	[22]

Continued

Table 1.6 Continued

Material and fabrication processes	Experimental test techniques and parameters	Main results	Study/ reference
Co-29Cr-6Mo alloy and Ti-6Al-4V	**Fretting apparatus and a reciprocating** sliding tribometer: Quasi-body fluid, Hanks's solution **Normal load**: 5 N **Frequency**: 10 Hz **Temperature** in the solutions: 37 ± 2°C	Co alloy exhibited good wear resistance; Ti alloy exhibited good fretting resistance	[47]
ASTM F1537 Co-Cr alloy	**Pin-on-disc** tribometer **Load**: 20 N **Rotation speed**: 60 rpm	The tribocorrosion properties of the Co-Cr alloy were enhanced by a layer of the S-phase	[48]
Co-Cr-Mo forged	**Pin-on-disc Load:** 9.8 N **Rotation speed**: 24 rpm	Forged CoCr exhibited a lower wear loss than a cast CoCr alloy	[49]
CoCr, stainless steel (SS), Al_2O_3, UHMWPE	**Pin-on-disc sliding speed**: 0.5 mm/s, 5 mm wear track radius **Normal load**: 1.8 N	UHMWPE shows the lowest friction and best wear performance against cartilage	[50]
AISI 316L CoCr29Mo6	**Pin-on-disc** for dry sliding wear tests **Load**: 5 N **Relative velocity**: 0.1 m/s **Ambient temperature:** 25°C	Ni-free high-nitrogen steel and LC-CoCrMo alloy exhibited higher wear resistance and dry friction than Ni-containing austenitic steels	[51]
Co-Cr-Mo low and high carbon	**Tribocorrosion techniques** **Load**: 1.2 N **Frequency**: 1 Hz **Temperature**: 37 ± 0.1°C Simulated body fluids [NaCl and phosphate-buffered solutions (PBS) with and without albumin]	LC-CoCrMo had a higher wear resistance in NaCl and PBS albumin than HC. No differences were observed for the alloys in the other solutions	[52]
Co-Cr alloy with boron additions (0, 0.3, 0.6, and 1 B wt%) by casting method	Three-axial hip joint simulator	Wear resistance as the boron increased	[53]
Commercial Ti-13Nb-3Zr alloy oxygen implanted	**Reciprocating type wear tester** **Normal forces**: 3, 5, and 10 N **Stroke length:** 10 mm and an alumina ball of 6 mm diameter was used as the counter surface	The implanted samples display a lower friction coefficient as compared to the substrate one	[54]

Table 1.6 Continued

Material and fabrication processes	Experimental test techniques and parameters	Main results	Study/ reference
Commercially pure titanium (CP-Ti) parts produced using selective laser melting (SLM) and casting	**Pin-on-disc** at room temperature: stainless steel disc of 45 mm diameter **Loads:** 15, 20, 25, and 30 N **Sliding speed:** 0.5 m/s for 15 min	SLM CP-Ti showed better wear resistance compared to casting due to fine grains and higher microhardness	[55]

(b) Bioactive: Forms direct chemical bonds with bone or even with soft tissue of a living organism, e.g., bioglass, glass ceramics, etc.
(c) Bioresorbable (biodegradable): Actively participates in the metabolic processes of an organism with predictable results, e.g., calcium phosphate ceramics, etc.

Zirconia is used as biomedical implants for its wear resistance. It has high mechanical strength and fracture toughness. However, MgO, CaO, Al_2O_3, and other oxides can be diffused in the material to improve the otherwise poor mechanical characteristics of pure zirconia [33]. Some significant research studies on zirconia, including developing stabilized zirconia, inducing bioactivity by chemical treatments, and improving the surface hardness, have been conducted. Silicon nitride (Si_3N_4) and silicon carbide (SiC) are also two kinds of biomedical materials frequently used because of their performance.

1.8.3 Polymeric biomedical materials

Metallic and ceramic materials have been broadly applied in biomedical engineering in the form of implants and devices because of their excellent mechanical properties, such as high moduli and stiffness [33]. However, the metallic ion generated from dissolution of metallic materials in body fluids can lead to allergies, poisoning, cellular reactions, and so on, which are often detrimental to human health and result in failure of the implant. In addition, many metallic materials have higher moduli than human bones; thus they can generate stress-shielding effects and result in loosening of implants. Ceramic materials also have some disadvantages, such as low fracture toughness. Therefore, polymer-based biomaterial becomes a promising alternative to metallic and ceramic material in biomedicine. Polymer-based materials when compared to metals and ceramics have many advantages, such as cost-efficiency, easy preparation, self-lubrication, and so on, and they are often used in the form of composites with additives to improve their tribological and mechanical behaviors. Nowadays, nanomaterial has attracted the attention of scientists in many domains because of its great value in basic research and potential industrial applications. Introducing nanomaterial into polymers can also enhance their mechanical and tribological performances greatly, due to the fine properties of nanomaterial such as small size, large surface areas, and higher activity. Bionic techniques have inspired many creative inventions in various spheres, such as super hydrophobic surfaces, dry adhesives, and so on, and scientists will continue to discover new biomimetic methods for use in basic research and industrial applications. The bionic research related to the tribology of polymer-based composites used in biomedicine has already begun, but it will take some time to make considerable progress. The following paragraphs describe the tribological behaviors of several polymer-based composites, including poly ether ether ketone (PEEK), epoxy resin (EP), ultrahigh molecular weight polyethylene (UHMWPE), and hydrogels in biomedicine.

1.8.3.1 UHMWPE as bearing material

UHMWPE has been used as a bearing surface in total joint prostheses for more than 45 years [56]. The majority of the joint replacement procedures performed around the world incorporate UHMWPE. Although some instances of metal-to-metal contact joints are found in the literature, still the most popular contact in artificial joints is between metal (hard) and plastic (soft). Metal acts as a substitute for the bone surface while the plastic is a substitute for the soft cartilage in the joints. Currently UHMWPE is the most widely used plastic in artificial joints because of its outstanding combination of physical and mechanical properties. It is most notable for its chemical inertness, lubricity, impact resistance, and abrasion resistance properties [56]. Some of the applications of UHMWPE in biomedical implants are shown in Fig. 1.14.

Tribological characterization of UHMWPE is relevant and a lot of work has been done on tribological testing of this material. The wear resistance of UHMWPE is found to be far better when compared to PTFE (polytetrafluoroethylene). UHMWPE is also found to exhibit a low dynamic friction coefficient (mean value of 0.1) under dry conditions with alumina as the counterface [57]. Low applied loads lead to a longer running-in time for UHMWPE. Wilches et al. [58] developed a modified pin-on-disc set-up in order to study the friction and wear behavior of stainless steel and Ti based alloy sliding against UHMWPE. It was found that adhesion of the polymer to the metallic surfaces is the most important wear mechanism in the case of metal-polymer pairs. The evidence of adhesive layers of UHMWPE can be clearly seen from Fig. 1.15. It is further interesting to see that, under dry conditions, stainless steel exhibits better friction and wear performance against UHMWPE when compared to Ti-based alloys [59]. Friction is greatly reduced by the presence of UHMWPE and this is believed to be due to the formation of a lubricating film of UHMWPE in the contact zone.

Biolubricants, namely sesame oil and nigella sativa oil, result in better friction and wear behavior (shown in Fig. 1.16) from UHMWPE as opposed to stainless steel, when compared to saline solution [60]. The decrease of friction and wear loss with oils as proposed by the authors might be attributed to fatty acids, which are the main component in any vegetable oil.

Developments in UHMWPE

Although the soft-on-hard contact combination is preferred in joint arthroplasty, the wear debris generated from the UHMWPE bearing surface is a source of many

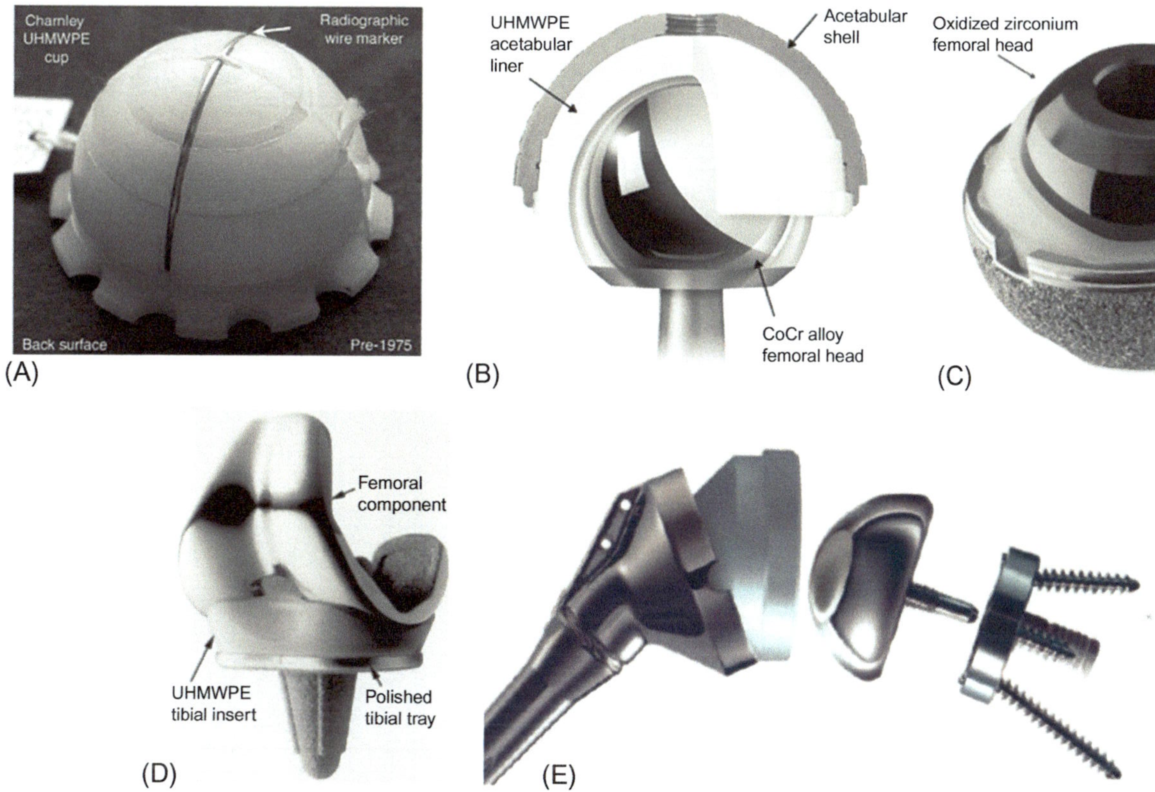

Fig. 1.14 Application of UHMWPE as biomedical implants: (A) cup; (B)–(D) liners; (E) total knee components; (F) shoulder prosthesis system components.
(A) From S.M. Kurtz, UHMWPE Biomaterials Handbook, second ed., Elsevier, London, 2009. (B) Courtesy of Stryker Orthopedics (Mahwah, New Jersey, USA). (C) Used with permission from Smith & Nephew Inc. (D) Used with permission from DePuy Synthes. (E) From http://morphopedics.wdfiles.com/local--files/reverse-total-shoulder-replacement/Prosthetic%20Components.jpg.

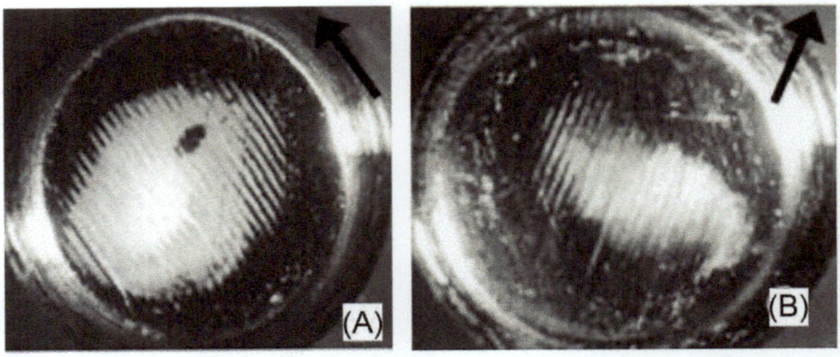

Fig. 1.15 Evidence of adhesion of UHMWPE layer on (A) Stainless steel and (B) Ti based alloy. The polymer layers correspond to the *whiter areas* in the image and the *black arrows* indicate the sliding direction [58].

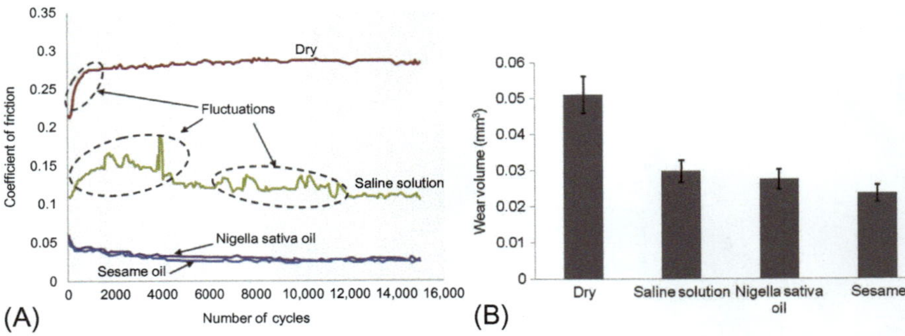

Fig. 1.16 (A) Friction and (B) Wear behavior of UHMWPE vs stainless steel under dry and lubricated conditions [60].

problems, such as aseptic loosening of the joint, initiation of an inflammatory reaction ultimately leading to osteolysis (pathological destruction of bone tissue), etc. For example, the friction of a stainless-steel head against UHMWPE typically produces a 0.2 mm/year linear wear rate [61]. Thus improving the wear resistance of the UHMWPE bearing surface is essential and there have many been efforts to do this. Massin and Achour [62] have reported the development of highly cross-linked UHMWPEs. Now, the abrasive wear resistance of the material depends on its degree of cross-linking. Its cross-linking rate is observed to increase proportionally to the irradiation doses, improving its wear resistance. However, cross-linking leads to a deterioration in the mechanical properties of UHMWPE and also makes the material more sensitive towards oxidation. A thermal treatment of the polymer is found to eliminate this problem. It is found that the addition of vitamin E (a powerful antioxidant) and other antioxidants also considerably increases wear resistance of the conventional UHMWPE and improves its clinical outcome. Moreover, the hard counterface and its surface condition also play a significant role in the wear of the polymer. Suitable surface coatings are applied on the counterface surface to maintain a low level of wear. But the durability of these coatings over longer periods is an issue. Harder materials,

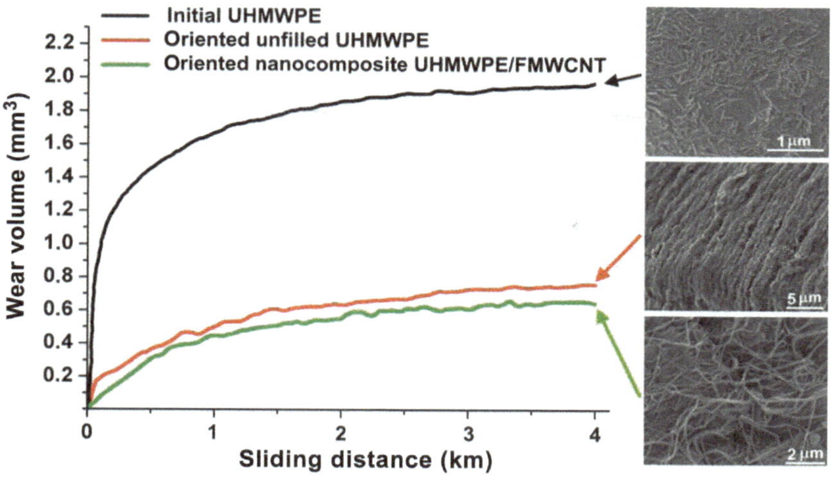

Fig. 1.17 Wear volume evolution for UHMWPE-based biomimetic materials. Initial UHMWPE has a lamellar structure, oriented unfilled UHMWPE has an oriented structure, and oriented nanocomposite UHMWPE/FMWCNT has a nanofibrillar structure [63].

such as ceramics, have also been used as the counterface with improved wear performance. Maksimkin et al. [63] report the results of pin-on-disc friction and wear testing for canine shoulder and knee joint cartilages and UHMWPE-based biomimetic composites against a stainless-steel counter-body. They found that orientation treatment and the incorporation of fluorinated carbon nanotubes applied to isotropic bulk UHMWPE caused both the reduction of the friction coefficient as well as the decrease of wear rate. Wear rate is found to decrease by about three times compared to initial UHMWPE (shown in Fig. 1.17).

1.8.3.2 Hydrogels as biomedical material

A hydrogel is a polymeric material that exhibits the ability to swell and retain a significant fraction of water within its structure, but will not dissolve in water [64]. Due to their high water content, porosity and soft consistency, hydrogels closely simulate natural living tissue, more so than any other class of synthetic biomaterials [65]. Hydrogels are used for manufacturing contact lenses, hygiene products, and wound dressings. Other commercial uses of hydrogels include applications in drug delivery and tissue engineering.

The prolonged life of natural joints is due to the presence of articular cartilage containing copious amounts of water at the joint interface. This enables the joint to operate at a lower coefficient of friction (0.02–0.002) and lower wear rate and hence have a longer life compared to artificial joints. Hydrogels such as poly vinyl alcohol (PVA) and poly (2-hydroxyethyl) methacrylate (polyHEMA) have shown the potential to substitute for natural cartilage and repair due to their cartilage-like structure and properties. They can also realize boundary lubrication, hydrodynamic lubrication, and separation of a sliding pair due to the squeezing of water under load.

The main concern about the use of PVA hydrogel in biomedical application is its poor mechanical properties, namely low compressive strength, low tear and shear resistance, and low creep resistance. A lot of research has been carried out to develop methods to enhance these properties. It is confirmed that the mechanical properties of PVA hydrogel can be improved by the procedure of repeated freezing-thawing, dehydrating in vacuum, and cross-linking by irradiation [66]. It has been reported that combining physical cross-linking and gamma irradiation cross-linking can produce chemical cross-linked PVA, which has a lower friction coefficient due to the condensed structure; also the water content decreases with the increase of cycles of freezing-thawing, the dose of irradiation, and the content of PVA [33]. Annealing is found to affect the mechanical properties of PVA in a positive way, but this method can lead to the collapse of pores, thus inducing a decrease in water content and lubricity. However, introducing polyethylene glycol [67] and acrylamide [68] into PVA during annealing can yield both good mechanical properties and lubricity. Introducing poly(vinyl pyrrolidone) (PVP) into PVA hydrogel can enhance the mechanical properties and decrease the friction coefficient due to the formation of a hydrogen bond [69]. It has also been further established that high wear and shear resistance of the hydrogel are contributed from the hydrophobic component [33].

When tested under lubrication it is observed that PVP-reinforced PVA hydrogel displays a lower coefficient of friction under synovial fluid compared to serum lubrication due to the high protein level of the latter, which results in high shear strength [70]. The tribological properties of PVA hydrogel are also influenced by many other factors such as lubricant, counterface, and degree of polymerization and saponification value [33]. PVP-reinforced PVA hydrogel has a lower friction coefficient under synovial fluid lubrication than serum lubrication due to the high protein level of serum, which induces high shear strength [70]. Nanohydroxyapatite reinforced PVA shows excellent osseointegration and in addition has high mechanical performance and a lower friction coefficient [71].

Table 1.7 shows a summary of various biomedical materials, their applications, and the major tribological properties possessed by them. A comparison of the wear rate of different material pairs is shown in Table 1.8.

1.9 Biofunctional coatings on biomaterials

Due to the special nature of the human body, implants and other biomaterials are often not compatible with living tissue due to the lack of biofunctionalities like blood compatibility and bioactivity. These may result in the development of infections after the surgical procedures. Often these infections cause device failure and need early replacement. Hence, in many cases the implant materials require surface modification before implantation. Again, many materials are structurally stable and biocompatible, but are prone to wear. For these materials, surface modification may render them suitable for broader applications.

For biomedical applications, controlling the surface and interfacial properties on the nanoscale is significant, since most of the biological reactions and host response

Table 1.7 Summary of tribology-based biomaterials [21]

Material	Application	Major properties
Alloy: titanium alloys, titanium aluminum vanadium alloy, cobalt chromium alloy, cobalt chromium molybdenum alloy	Total joint replacement	Wear and corrosion resistance
Inorganic: diamond-like carbon	Biocompatible coatings	Reduced friction and increased wear resistance
Ceramics: Al_2O_3, ZrO_2, Si_3N_4, SiC, B_4C, quartz, bioglass (Na_2O-CaO-SiO_2-P_2O_5), sintered hydroxyapatite ($Ca_{10}(PO_4)6(OH)_2$)	Bone joint coating	Wear and corrosion resistance
Polymers: ultrahigh molecular weight polyethylene, polytetrafluoroethylene (PTFE), poly(glycolic acid)	Joint socket interpositional implant temporomandibular joint (jaw) joint bone	Wear and corrosion resistance, low friction coefficient, elastics with less wear
Composites: specialized silicon polymers	Bone joint	Wear, corrosion, and fatigue resistance

From B. Shi, H. Liang, Tribological applications of biomaterials: an overview, Sci. China (Ser. A) 44 (2001). Used with permission from the Science China Press.

Table 1.8 Comparison of volumetric wear rates for various artificial hip joints tested in simulators [72]

Material combination	Volumetric wear (mm^3/million cycles)
UMWPE-on-metal	40
UHMWPE-on-ceramic	25
Cross-linked UHMWPE-on-metal	5–10
Metal-on-metal	1.0
Ceramic-on-ceramic	0.1

happens at the surface of the substrate. Thus the structures and chemical assemblies of nanostructures and nanoassemblies have a direct impact on the macroscopic properties of the material. Self-assembled monolayers (SAMs) are nanosized coatings that offer a flexible method of carrying out surface modification of biomaterials [73], done to tailor surface properties of the material for specific end applications. These nanocoatings can serve primary functions such as cover the surface, provide etch protection and anticorrosion, along with a host of other secondary chemical functions such as drug delivery and biocompatibility. A variety of surface modification techniques are presently available due to advancements in nanoscience and nanotechnology. The surface modification techniques that are used for modifying metallic biomaterials can broadly be classified as mechanical, chemical, and physical surface modification methods [73].

1.9.1 Mechanical techniques

Mechanical techniques are the traditional methods of surface modification of materials. These are straightforward methods with no addition of materials. Instead the base material itself is machined in such a way as to yield the desired level of surface finish. Mechanical surface modification methods include machining, grinding, polishing, and blasting. It also includes physical treatment, shaping, or etching the material surface.

1.9.2 Chemical treatment

The chemical method of surface modification involves chemical reactions happening at the interface between the metal surface and a solution. A variety of chemical treatments are available based on the employed chemical pathways.

- Acid treatment produces clean and uniform surfaces.
- Hydrogen peroxide treatment induces the formation of the corresponding metal peroxy gels, which improve the bioactivity of metallic implants like titanium.
- Alkali and heat treatments result in the formation of a biologically active bone-like apatite layer on the surface of bioactive ceramics.
- Various types of sol-gel coatings are also an important class of methods under the chemical treatment processes.
- Chemical vapor deposition (CVD) is also used for modifying the mechanical and biological properties of metal alloys. CVD is a technique that involves the formation of a nonvolatile layer on the substrate by the chemical reaction of the gas phase along with the substrate surface. CVD is useful especially in coating complex geometries.

Apart from these processes, there are several other coatings that find their uses in biomedical applications. Ceramic coatings deposited on metals help in improving the bone bonding ability for orthopedic applications. Calcium phosphate coatings (hydroxyapatite), for example, are frequently used in orthopedic applications to enhance bone integration. Silica gel is also employed for the formation of a bone-like apatite on the surface. Oxide films are formed on the metal surface by anodic oxidation, which is a process whereby the reactions at the electrodes in an electrolyte are controlled by an externally applied electric field. The advantages of this method are improved adhesion and bonding along with increased resistance against corrosion and wear. Biochemical modification of metal alloys involves understanding the biological and biochemical processes of cell function, adhesion, differentiation, and remodeling. In short, the process enables the induction of specific cell and tissue responses by using surface immobilized molecules, such as peptides, proteins, or growth factors. A variety of techniques have been used with this objective in mind, which include silanized titania, photochemistry, and SAMs.

1.9.3 Physical methods

These methods do not involve the use of chemical reactions. Physical vapor deposition (PVD), thermal spraying methods, glow discharge plasma treatment, and ion implantation techniques are some of the popular physical methods of surface treatments.

- PVD is characterized by a process in which the material goes from a condensed phase to a vapor phase and then back to a thin film condensed phase. The advantages of this method are high coating density, strong adhesion, multicomponent layers, low substrate temperature, and the choice of a variety of substrate and coating materials. The most common PVD processes are sputtering, evaporation, and ion plating. Titanium based coatings viz. TiC and TiN are deposited by evaporating Ti in C_2H_2 (acetylene) and N_2 plasma, respectively.
- In thermal spraying methods, the coating material that is present in powder or wire form is melted into tiny droplets and sprayed onto surfaces at high velocity. The individual particles stick and condense on the substrate. The coating is formed by continuous buildup of successive layers of condensed droplets. Plasma spraying, flame spraying, arc spraying, detonation gun spraying, laser spraying, and high velocity oxy-fuel (HVOF) spraying are popular thermal spraying techniques. Plasma spraying is used in the formation of hydroxyapatite coatings on endoprostheses. Due to this, the implant forms a strong connection with the bone tissue [74]. Both calcium silicate coatings and titanium coatings reported to possess good bioactivity and biocompatibility are developed by spraying techniques [75]. Plasma sprayed titanium coating, having a porous structure, has found usage in teeth, root, hip, knee, and shoulder implants [76].
- Ion implantation is a low-temperature process in which ions of one elements are accelerated and bombarded into a solid target (substrate) thereby changing the physical, chemical, or electrical properties of the substrate surface. The source of the energetic particles is plasma, a compound or alloy sputtering target, vacuum/plasma arcs, or special ion sources. Ion plating has been used in cardiovascular applications [73]. Dental prosthesis is another application of the ion plating process, especially when Co-CR-Mo alloy is involved [77].
- Deposition by the sputtering technique is also popular due to its simplicity, versatility, and flexibility. This process involves ejecting material from a "target" that is a source onto a "substrate" such as a silicon wafer. Ion beam sputtering is a sophisticated method that provides better control of the deposition process [78].

Other methods include enhanced plasma ionization, which can be achieved by either additional gas ionization or plasma confinement. Glow discharge plasma is a type of surface plasma treatment that involves low temperature and a low pressure gas in which ionization is controlled by energetic electrons. Glow discharge plasma is already an established surface processing technique in the microelectronics industry. The process has also found use in surface modification of bulk polymers and making thin polymer coatings, which is important in the biomaterials research domain [79]. Plasma treatment has also been the preferred method for increasing the surface energy and cleaning the surface of biomaterials before biological evaluation studies [80]. Investigations have been conducted on the suitability of glow discharge plasma from the perspective of surface cleaning and modification of titanium-based metallic biomaterials through plasma treatment of the component in an atmosphere containing argon and oxygen [81]. Sobiecki et al. [82] examined the influence of glow discharge nitriding, oxynitriding, and carbonitriding on the surface modification of Ti-1Mg-1Mn alloy. These treatments produce surface layers with a diffusion character exhibiting high hardness and good wear and corrosion resistance as well as increased fatigue limit. The ion beam implantation technique is also employed for surface modification of biomaterials [83]. In this process energetic ions are introduced into the surface layer of a solid substrate by bombardment. Wear resistance and bone conductivity can be improved by nitrogen and calcium ion implantation, respectively. Ion beam implantation has been shown to improve the corrosion resistance of metals and alloys. Phosphorus ion implantation is found to increase the corrosion resistance of titanium, which is popular as an implant material [84].

1.10 Tribology of medical devices and surgical instruments

Many medical devices are now regularly used in the treatment of humans. They can be as elementary as a simple bandage and can also be as complex as imaging equipment. Common examples of medical devices include instruments, apparatus, appliances, materials, and so forth intended to be used in human beings for the purposes of diagnosis, prevention, monitoring, treatment, or alleviation of disease; compensation for an injury or handicap; or investigation, replacement, or modification of the anatomy or of a physiological process [2]. Because of the nature of their applications, medical devices are heavily regulated. Strict and comprehensive preclinical testing has become crucial in the evaluation of new innovative medical devices. Since many medical devices involve relative moving parts, friction and wear not only affect the functioning of these devices but also have the potential to adversely affect the natural tissues.

The most common medical devices that actively involve biotribology are those based on the musculoskeletal, dental, and cardiovascular systems. In addition, a host of other systems, such as ocular and skin-based systems, are active examples of biotribology. These devices are either in contact with native tissues or with the biomaterials, and are often under structural loading. Tribological issues related to friction and wear on the moving parts not only affect the function of these devices but may also have adverse effects on the natural tissues. Hence, biotribology plays an important role in many medical devices.

1.11 Closure

The field of biomaterials is increasing in importance due to the high demands of an aging population as well as the increasing average weight of people [38]. Biomedical engineering deals with various implants and prosthetics that are used in the living human body. With the increase in the complexities of human life, there has been a global rise in prosthetics use, especially joint replacements. Hence, detailed study of tribological aspects of the materials used, namely friction and wear, is a vital necessity. Moreover, as many more younger people are now being subjected to joint replacement procedures, research in the field of tribology has become significant in order to push the limits of these materials in terms of their efficiency as well as service life. Researchers are on a constant look-out for the ideal pair of arthroplasty materials to minimize wear. Moreover, many surface modification techniques suitable for biomedical applications have been and are being developed. Although many advances have been recently seen in this field, a great deal of scope still exists for further development and improvements.

References

[1] M.S. Scholz, J.P. Blanchfield, L.D. Bloom, B.H. Coburn, M. Elkington, J.D. Fuller, et al., The use of composite materials in modern orthopaedic medicine and prosthetic devices: a review, Compos. Sci. Technol. 71 (16) (2011) 1791–1803.

[2] Z.M. Jin, J. Zheng, W. Li, Z.R. Zhou, Tribology of medical devices, Biosurf. Biotribol. 2 (4) (2016) 173–192.

[3] M. Hussein, A. Mohammed, N. Al-Aqeeli, Wear characteristics of metallic biomaterials: a review, Materials 8 (5) (2015) 2749.

[4] D.C. Hansen, Metal corrosion in the human body: the ultimate bio-corrosion scenario, Electrochem. Soc. Interface 17 (2) (2008) 31–34.

[5] Z.R. Zhou, Z.M. Jin, Biotribology: recent progresses and future perspectives, Biosurf. Biotribol. 1 (1) (2015) 3–24.

[6] L.S.J. Berrien, Biotribology: Studies of the Effects of Biochemical Environments on the Wear and Damage of Articular Cartilage, Faculty of Virginia Polytechnic Institute and State University, Blacksburg, VA, 1999.

[7] V.C. Mow, G.A. Ateshian, R.L. Spilker, Biomechanics of diarthrodial joints: a review of twenty years of progress, J. Biomech. Eng. 115 (4b) (1993) 460–467.

[8] M. Geetha, A.K. Singh, R. Asokamani, A.K. Gogia, Ti based biomaterials, the ultimate choice for orthopaedic implants—a review, Prog. Mater. Sci. 54 (3) (2009) 397–425.

[9] P. Massin, E. Viguier, B. Flautre, P. Hardouin, E. Astoin, B. Duponchel, Migration of polyethylene debris along well-fixed cemented implants, *J. Biomed. Mater. Res.* B Appl. Biomater. 68 (2) (2004) 140–148.

[10] M.F. Baslé, G. Bertrand, S. Guyetant, D. Chappard, M. Lesourd, Migration of metal and polyethylene particles from articular prostheses may generate lymphadenopathy with histiocytosis, J. Biomed. Mater. Res. 30 (2) (1996) 157–164.

[11] E. Ingham, J. Fisher, Biological reactions to wear debris in total joint replacement, Proc. Inst. Mech. Eng. H 214 (1) (2000) 21–37.

[12] H.A. McKellop, P. Campbell, S.H. Park, T.P. Schmalzried, P. Grigoris, H.C. Amstutz, et al., The origin of submicron polyethylene wear debris in total hip arthroplasty, Clin. Orthop. Relat. Res. 311 (1995) 3–20.

[13] H.P. Sieber, C.B. Rieker, P. Kottig, Analysis of 118 second-generation metal-on-metal retrieved hip implants, J. Bone Joint Surg. 81 (1) (1999) 46–50.

[14] J.B. Medley, F.W. Chan, J.J. Krygier, J.D. Bobyn, Comparison of alloys and designs in a hip simulator study of metal on metal implants, Clin. Orthop. Relat. Res. 329 (Suppl) (1996) S148–S159.

[15] J.P. Simon, G. Fabry, An overview of implant materials, Acta Orthop. Belg. 57 (1) (1991) 1–5.

[16] J. Charnley, Z. Cupic, The nine and ten year results of the low-friction arthroplasty of the hip, Clin. Orthop. Relat. Res. 95 (1973) 9–25.

[17] Y. Luo, L. Yang, M. Tian, Application of biomedical-grade titanium alloys in trabecular bone and artificial joints, in: J.P. Davim (Ed.), Biomaterials and Medical Tribology, Woodhead Publishing, Cambridge, UK, 2013, pp. 181–216.

[18] M. Fini, G. Giavaresi, P. Torricelli, A. Krajewski, A. Ravaglioli, M.M. Belmonte, et al., Biocompatibility and osseointegration in osteoporotic bone, J. Bone Joint Surg. 83 (1) (2001) 139–143.

[19] K. Suzuki, K. Aoki, K. Ohya, Effects of surface roughness of titanium implants on bone remodeling activity of femur in rabbits, Bone 21 (6) (1997) 507–514.

[20] C. Zietz, C. Fabry, J. Reinders, R. Dammer, J.P. Kretzer, R. Bader, et al., Wear testing of total hip replacements under severe conditions, Expert Rev. Med. Devices 12 (4) (2015) 393–410.

[21] F. Zivic, M. Babic, N. Grujovic, S. Mitrovic, Tribometry of materials for bioengineering applications, Tribol. Ind. 32 (1) (2010) 25–32.

[22] A. Choubey, B. Basu, R. Balasubramaniam, Tribological behaviour of Ti-based alloys in simulated body fluid solution at fretting contacts, Mater. Sci. Eng. A 379 (1–2) (2004) 234–239.

[23] S.S. Brown, I.C. Clarke, A review of lubrication conditions for wear simulation in artificial hip replacements, Tribol. Trans. 49 (1) (2006) 72–78.

[24] V. Saikko, P. Paavolainen, M. Kleimola, P. Statis, A five-station hip joint simulator for wear rate studies, Proc. Inst. Mech. Eng. H 206 (4) (1992) 195–200.

[25] C.R. Bragdon, D.O. O'Connor, J.D. Lowenstein, M. Jasty, W.D. Syniuta, The importance of multidirectional motion on the wear of polyethylene, Proc. Inst. Mech. Eng. H 210 (3) (1996) 157–165.

[26] J.P. Paul, Paper 8: Forces transmitted by joints in the human body, Proc. Instit. Mech. Eng. Conf. Proc. 181 (10) (1966) 8–15.

[27] T.P. Schmalzried, E.F. Shepherd, F.J. Dorey, W.O. Jackson, M. dela Rosa, F. Fa'vae, et al., The John Charnley Award. Wear is a function of use, not time, Clin. Orthop. Relat. Res. 381 (2000) 36–46.

[28] R.M. Streicher, Examination of the Tribological Behaviour, Sulzer Medical Technology Ltd., Winterthur, 1997.

[29] I.C. Clarke, F.W. Chan, A. Essner, V. Good, C. Kaddick, R. Lappalainen, et al., Multi-laboratory simulator studies on effects of serum proteins on PTFE cup wear, Wear 250 (1–12) (2001) 188–198.

[30] Z. Lu, H. McKellop, Frictional heating of bearing materials tested in a hip joint wear simulator, Proc. Inst. Mech. Eng. H 211 (1) (1997) 101–108.

[31] H. McKellop, I.C. Clarke, K.L. Markolf, H.C. Amstutz, Wear characteristics of UHMW polyethylene: a method for accurately measuring extremely low wear rates, J. Biomed. Mater. Res. 12 (6) (1978) 895–927.

[32] A. Wang, A. Essner, C. Stark, J.H. Dumbleton, Comparison of the size and morphology of UHMWPE wear debris produced by a hip joint simulator under serum and water lubricated conditions, Biomaterials 17 (9) (1996) 865–871.

[33] E. S. F, L. Shi, Z.G. Guo, W.M. Liu, The recent progress of tribological biomaterials, Biosurf. Biotribol. 1 (2) (2015) 81–97.

[34] M.Z. Ibrahim, A.A.D. Sarhan, F. Yusuf, M. Hamdi, Biomedical materials and techniques to improve the tribological, mechanical and biomedical properties of orthopedic implants—A review article, J. Alloys Compd. 714 (2017) 636–667.

[35] M. Niinomi, Recent metallic materials for biomedical applications, Metall. Mater. Trans. A 33A (2002) 477–486.

[36] M. Niinomi, Metallic biomaterials, J. Artif. Organs 11 (3) (2008) 105–110.

[37] T. Hanawa, Metal ion release from metal implants, Mater. Sci. Eng. C 24 (6–8) (2004) 745–752.

[38] V.S. Viteri, E. Fuentes, Titanium and titanium alloys as biomaterials, in: J. Gegner (Ed.), Tribology—Fundamentals and Advancements, InTech, Rijeka, 2013. (pp. Ch. 05).

[39] B. Stadlinger, S.J. Ferguson, U. Eckelt, R. Mai, A.T. Lode, R. Loukota, et al., Biomechanical evaluation of a titanium implant surface conditioned by a hydroxide ion solution, Br. J. Oral Maxillofac. Surg. 50 (1) (2012) 74–79.

[40] Y.L. Zhou, M. Niinomi, T. Akahori, H. Fukui, H. Toda, Corrosion resistance and biocompatibility of Ti–Ta alloys for biomedical applications, Mater. Sci. Eng. A 398 (1–2) (2005) 28–36.

[41] B.D. Ratner, A.S. Hoffman, F.J. Schoen, J.E. Lemons, Biomaterials Science: An Introduction to Materials in Medicine, Academic Press, San Diego, CA, 1996.

[42] I. Cvijović-Alagić, Z. Cvijović, S. Mitrović, M. Rakin, Đ. Veljović, M. Babić, Tribological behaviour of orthopaedic Ti-13Nb-13Zr and Ti-6Al-4V alloys, Tribol. Lett. 40 (1) (2010) 59–70.
[43] S. Gialanella, G. Ischia, G. Straffelini, Phase composition and wear behavior of NiTi alloys, J. Mater. Sci. 43 (5) (2008) 1701–1710.
[44] K.S. Suresh, M. Geetha, C. Richard, J. Landoulsi, H. Ramasawmy, S. Suwas, et al., Effect of equal channel angular extrusion on wear and corrosion behavior of the orthopedic Ti–13Nb–13Zr alloy in simulated body fluid, Mater. Sci. Eng. C 32 (4) (2012) 763–771.
[45] F. Mamoun, A. Omar, L. Mohamed, D. Leila, A. Iost, in: Comparative study on tribological behavior of Ti-6Al-7Nb and SS AISI 316L alloys, for total hip prosthesis, TMS 2014: 143rd Annual Meeting & Exhibition: Annual Meeting Supplemental Proceedings, Springer International Publishing, Cham, 2016, pp. 237–246.
[46] S.J. Li, R. Yang, S. Li, Y.L. Hao, Y.Y. Cui, M. Niinomi, et al., Wear characteristics of Ti–Nb–Ta–Zr and Ti–6Al–4V alloys for biomedical applications, Wear 257 (9) (2004) 869–876.
[47] A. Iwabuchi, J.W. Lee, M. Uchidate, Synergistic effect of fretting wear and sliding wear of Co-alloy and Ti-alloy in Hanks' solution, Wear 263 (1) (2007) 492–500.
[48] X. Luo, X. Li, Y. Sun, H. Dong, Tribocorrosion behavior of S-phase surface engineered medical grade Co–Cr alloy, Wear 302 (1) (2013) 1615–1623.
[49] A. Chiba, K. Kumagai, N. Nomura, S. Miyakawa, Pin-on-disk wear behavior in a like-on-like configuration in a biological environment of high carbon cast and low carbon forged Co–29Cr–6Mo alloys, Acta Mater. 55 (4) (2007) 1309–1318.
[50] S.M.T. Chan, C.P. Neu, K. Komvopoulos, A.H. Reddi, P.E. Di Cesare, Friction and wear of hemiarthroplasty biomaterials in reciprocating sliding contact with articular cartilage, J. Tribol. *133* (4) (2011). 041201-1-7.
[51] A. Fischer, S. Weiß, M.A. Wimmer, The tribological difference between biomedical steels and CoCrMo-alloys, J. Mech. Behav. Biomed. Mater. 9 (2012) 50–62.
[52] A. Igual Muñoz, S. Mischler, Effect of the environment on wear ranking and corrosion of biomedical CoCrMo alloys, J. Mater. Sci. Mater. Med. 22 (3) (2011) 437–450.
[53] M. Alvarez-Vera, J.A. Ortega-Saenz, M.A.L. Hernandez-Rodríguez, A study of the wear performance in a hip simulator of a metal–metal Co–Cr alloy with different boron additions, Wear 301 (1) (2013) 175–181.
[54] L. Mohan, C. Anandan, Wear and corrosion behavior of oxygen implanted biomedical titanium alloy Ti–13Nb–13Zr, Appl. Surf. Sci. 282 (2013) 281–290.
[55] H. Attar, K.G. Prashanth, A.K. Chaubey, M. Calin, L.C. Zhang, S. Scudino, et al., Comparison of wear properties of commercially pure titanium prepared by selective laser melting and casting processes, Mater. Lett. 142 (2015) 38–41.
[56] S.M. Kurtz, UHMWPE Biomaterials Handbook, second ed., Elsevier, London, 2009.
[57] F. Zivic, M. Babic, S. Mitrovic, D. Adamovic, S. Pelemis, Friction coefficient of UHMWPE during dry reciprocating sliding, Tribol. Ind. 36 (3) (2014) 281–286.
[58] L.V. Wilches, J.A. Uribe, A. Toro, Wear of materials used for artificial joints in total hip replacements, Wear 265 (1–2) (2008) 143–149.
[59] A. Ruggiero, R. D'Amato, E. Gómez, Experimental analysis of tribological behavior of UHMWPE against AISI420C and against TiAl6V4 alloy under dry and lubricated conditions, Tribol. Int. 92 (2015) 154–161.
[60] M. Guezmil, W. Bensalah, S. Mezlini, Effect of bio-lubrication on the tribological behavior of UHMWPE against M30NW stainless steel, Tribol. Int. 94 (2016) 550–559.
[61] J. Charnley, D.K. Halley, Rate of wear in total hip replacement, Clin. Orthop. Relat. Res. 112 (1975) 170–179.

[62] P. Massin, S. Achour, Wear products of total hip arthroplasty: the case of polyethylene, Morphologie 101 (332) (2017) 1–8.
[63] A.V. Maksimkin, D.I. Chukov, F.S. Senatov, A.I. Salimon, Comparative analysis of the tribological characteristics of canine joint cartilage and UHMWPE-based biomimetic materials, Mater. Lett. 191 (2017) 105–107.
[64] E.M. Ahmed, Hydrogel: Preparation, characterization, and applications: a review, J. Adv. Res. 6 (2) (2015) 105–121.
[65] E. Caló, V.V. Khutoryanskiy, Biomedical applications of hydrogels: a review of patents and commercial products, Eur. Polym. J. 65 (2015) 252–267.
[66] Z.Q. Gu, J.M. Xiao, X.H. Zhang, The development of artificial articular cartilage—PVA-hydrogel, Biomed. Mater. Eng. 8 (2) (1998) 75–81.
[67] H. Bodugoz-Senturk, J. Choi, E. Oral, J.H. Kung, C.E. Macias, G. Braithwaite, et al., The effect of polyethylene glycol on the stability of pores in polyvinyl alcohol hydrogels during annealing, Biomaterials 29 (2) (2008) 141–149.
[68] H. Bodugoz-Senturk, C.E. Macias, J.H. Kung, O.K. Muratoglu, Poly(vinyl alcohol)-acrylamide hydrogels as load-bearing cartilage substitute, Biomaterials 30 (4) (2009) 589–596.
[69] R. Ma, D. Xiong, F. Miao, J. Zhang, Y. Peng, Friction properties of novel PVP/PVA blend hydrogels as artificial cartilage, *J. Biomed. Mater. Res.* A 93 (3) (2010) 1016–1019.
[70] J.K. Katta, M. Marcolongo, A. Lowman, K.A. Mansmann, Friction and wear behavior of poly(vinyl alcohol)/poly(vinyl pyrrolidone) hydrogels for articular cartilage replacement, *J. Biomed. Mater. Res.* A 83 (2) (2007) 471–479.
[71] A.S. Maiolo, M.N. Amado, J.S. Gonzalez, V.A. Alvarez, Development and characterization of Poly (vinyl alcohol) based hydrogels for potential use as an articular cartilage replacement, *Mater. Sci. Eng.* C 32 (6) (2012) 1490–1495.
[72] Z.M. Jin, M. Stone, E. Ingham, J. Fisher, Biotribology, Curr. Orthop. *20* (1) (2006) 32–40.
[73] A. Mahapatro, Bio-functional nano-coatings on metallic biomaterials, Mater. Sci. Eng. C 55 (2015) 227–251.
[74] K. De Groot, R. Geesink, C.P.A.T. Klein, P. Serekian, Plasma sprayed coatings of hydroxylapatite, J. Biomed. Mater. Res. 21 (12) (1987) 1375–1381.
[75] T. Kokubo, Surface chemistry of bioactive glass-ceramics, J. Non Cryst. Solids 120 (1) (1990) 138–151.
[76] S. Vercaigne, J.G.C. Wolke, I. Naert, J.A. Jansen, Histomorphometrical and mechanical evaluation of titanium plasma-spray-coated implants placed in the cortical bone of goats, J. Biomed. Mater. Res. 41 (1) (1998) 41–48.
[77] O. Knotek, F. Löffler, K. Weitkamp, Physical vapour deposition coatings for dental prostheses, Surf. Coat. Technol. 54 (1992) 536–540.
[78] W.D. Sproul, Physical vapor deposition tool coatings, Surf. Coat. Technol. 81 (1) (1996) 1–7.
[79] M. Noeske, J. Degenhardt, S. Strudthoff, U. Lommatzsch, Plasma jet treatment of five polymers at atmospheric pressure: surface modifications and the relevance for adhesion, Int. J. Adhes. Adhes. 24 (2) (2004) 171–177.
[80] D.C. Smith, R.M. Pilliar, J.B. Metson, N.S. McIntyre, Dental implant materials. II. Preparative procedures and surface spectroscopic studies, J. Biomed. Mater. Res. 25 (9) (1991) 1069–1084.
[81] B.O. Aronsson, J. Lausmaa, B. Kasemo, Glow discharge plasma treatment for surface cleaning and modification of metallic biomaterials, J. Biomed. Mater. Res. 35 (1) (1997) 49–73.
[82] J.R. Sobiecki, T. Wierzchoń, J. Rudnicki, The influence of glow discharge nitriding, oxynitriding and carbonitriding on surface modification of Ti–1Al–1Mn titanium alloy, Vacuum 64 (1) (2001) 41–46.

[83] J. Jagielski, A. Piatkowska, P. Aubert, L. Thomé, A. Turos, A. Abdul Kader, Ion implantation for surface modification of biomaterials, Surf. Coat. Technol. 200 (22) (2006) 6355–6361.

[84] D. Krupa, J. Baszkiewicz, J.A. Kozubowski, A. Barcz, J.W. Sobczak, A. Biliński, et al., Effect of phosphorus-ion implantation on the corrosion resistance and biocompatibility of titanium, Biomaterials 23 (16) (2002) 3329–3340.

Further reading

[85] Americal Academy of Orthopaedic Surgeons, http://orthoinfo.aaos.org/topic.cfm?topic=a00377, 2015.

[86] Knee Replacement & Hip Replacement Patient Advocacy & Online Community, https://bonesmart.org/, 2017.

[87] J. Walczak, F. Shahgaldi, F. Heatley, In vivo corrosion of 316L stainless-steel hip implants: morphology and elemental compositions of corrosion products, Biomaterials 19 (1998) 229–237.

[88] B. Shi, H. Liang, Tribological applications of biomaterials: an overview, Sci. China (Series A) 44 (2001).

Designing and analysis of the femoral neck for an artificial hip joint prosthesis

2

Luke Hocking, A. Pramanik*, A.K. Basak†, S. Chattopadhyaya‡*
*School of Civil and Mechanical Engineering, Curtin University, Bentley, WA, Australia,
†Adelaide microscopy, The University of Adelaide, Adelaide, SA, Australia, ‡Department of Mech. Eng., IIT-ISM, Dhanbad, India

2.1 Introduction

The hip joint is the largest and one of the most important joints in the human body; it grants mobility to the human body via walking, running, and jumping. Over time the hip joint will experience wear and tear that may result in the articular cartilage surrounding the femur depleting, leading to osteoarthritis and causing the patient pain and lack of mobility. A common solution to this problem is to replace the damaged hip joint with an artificial one. A typical total hip replacement prosthesis consists of a stem and femoral neck inserted into the top of the femur, a femoral head attached to the femoral neck, an acetabular shell fitted to the hip bone, and an acetabular cup liner fixed to the inside of the acetabular cup and against the femoral head.

Although the typical metal-on-polyethylene (M-on-PE) prosthesis design is the most common, there are continuous advances in the technology of medical prostheses and hence new alternative methods and designs are always being tested. One of the most common problems that arises with an M-on-PE total hip replacement is polyethylene debris. Polyethylene debris "creates periprosthetic osteolysis by the release of cytokines and proteolytic enzymes—ultimately leading to implant failure" [1]. A metal-on-metal or M-on-M prosthesis consists of a similar design to M-on-PE, but with a metal acetabular cup instead of a polyethylene one. There is no polyethylene debris in this case, so these prostheses are not subject to the same common failures as the M-on-PE designs. Ceramic-on-ceramic (C-on-C) hip replacement prostheses comprise a ceramic femoral head that connects to a ceramic acetabular cup. This design was created to combat the problems of friction and wear in other prostheses [2]. C-on-C implants have a "high level of hardness, scratch resistance and the inert nature of debris" in comparison to both M-on-PE and M-on-M [1].

With increased life expectancy and a growing number of high-energy traumas, fractures of the femoral neck are becoming an increasingly frequent surgical entity [3, 4]. These fractures have bimodal characteristics: young patients, victims of high-energy trauma, and the elderly, which accounts for most of the injuries [5]. Fixation of femoral neck fractures is associated with a higher incidence of complications than any other fracture [6]. The femoral neck area is subjected to large compressive and sheer forces associated with ambulation. Even in the most sedentary individual, the daily cyclic

Mechanical Behavior of Biomaterials. https://doi.org/10.1016/B978-0-08-102174-3.00002-4

loading of the hip and femoral neck produces high stresses on the bony trabeculae in this anatomic region. In long-distance runners and other high-performance athletes, the forces across the femoral neck are multiplied exponentially because the athletes' training regimens place tremendous physical burdens on this relatively small bridge of bone, which connects the femoral head to the diaphysis (http://emedicine.medscape.com/article/1246691-overview). The research and analysis on this bone is lacking though it is imperatively needed to improve the artificial hip joint prosthesis. This study considers various designs for the femoral neck and analyzes the load-bearing capacity of each design. In addition, the effect of load application angles of stress generation has also been investigated. This investigation is specifically focused on the design and analysis of the femoral neck component of the prosthesis and not the entire stem design, so the geometry of existing neck designs is more closely examined. Finite element analysis is a powerful method that is extensively applied in the biomechanical study of implant performance. This method simulates in vivo mechanical conditions, and different configurations of implant design as well as other complex problems can be assessed and tested using a computer rather than in vitro destructive and mechanical testing [7, 8]

2.2 Design background

The most common configuration of hip replacement prosthesis used today is the metal-on-polyethylene, or M-on-PE. This consists of a metal stem, femoral neck, femoral head, and acetabular shell with an ultra-high molecular weight polyethylene (UHMWPE) acetabular cup lining. The metal can vary from stainless steel to cobalt chromium alloy, or Co-Cr, or a titanium alloy. Titanium alloys are widely used in orthopedics because of their high strength, low weight, low modulus of elasticity, high corrosion resistance, and excellent biocompatibility. There are a number of various titanium alloys, each of which have specific properties. Table 2.1 shows a list of different titanium alloys with their corresponding properties.

The first four alloys shown in Table 2.1, pure grades 1 through to 4, are all alloys that consist of pure titanium with a single alpha phase microstructure. These alloys have a very particular oxygen content of around 0.2%–0.5%, which increases the material's strength. Ductile titanium wires, which are usually used in fiber meshes, are manufactured from Grade 1 [10]. Grades 3 and 4 have higher tensile strengths and are generally used for acetabular cup shells with self-tapping threads. These grades, however, are not suitable for other components of the prosthesis, because the stem and neck are subject to much higher bending stresses. Ti-6Al-4V (T64) alloys have much stronger properties than pure titanium grades and have been developed for structural materials and are commonly used in medical applications [10]. Newer types of alloys are being produced that replace vanadium, V, with niobium, Nb, or iron, Fe, in an attempt to combat any potential toxicity and increase the strength of the alloy. These alloys are, however, still not as widely used as T64 and T64 ELI, which are mainly used for anchorage stems of femoral components with wear-resistant cobalt-chromium-molybdenum or Co-Cr-Mo metal or alumina ceramic ball heads [10]. Thus, typical materials for a metal-on-polyethylene hip prosthesis are T64 ELI for the stem

Table 2.1 Properties of titanium alloys for biomedical applications [9]

Alloy	Tensile strength (MPa)	Yield strength (σ_y)	Elongation	Reduction area (%)	Elastic modulus (GPa)	Phase
Pure grade 1	240	170	24	30	102.7	α
Pure grade 2	345	275	20	30	102.7	α
Pure grade 3	450	380	18	30	103.4	α
Pure grade 4	550	485	15	25	104.1	α
Ti-6Al-4V	895–930	825–869	6–10	20–25	110–114	$\alpha+\beta$
Ti-6Al-4V ELI	860–965	795–875	10–15	25–47	101–110	$\alpha+\beta$
Ti-6Al-7Nb	900–1050	880–950	8.1–15	25–45	114	$\alpha+\beta$
Ti-5Al-2.5Fe	1020	895	15	35	112	$\alpha+\beta$

and femoral neck, Co-Cr-Mo for the femoral head, UHMWPE for the acetabular liner cup, and pure Ti Grade 4 for the acetabular cup shell.

The geometry of hip replacement prostheses differs dramatically from patient to patient. This raises difficulties when trying to determine a typical set of dimensions to design around. The cross-sectional shape as well as the side profiles of the prosthesis can also differ greatly to provide room for creativity and refinement. Some examples of the different possibilities in the hip replacement prosthesis design as well as some example lengths are shown in Figs. 2.1 and 2.2. Fig. 2.1 shows the combination of

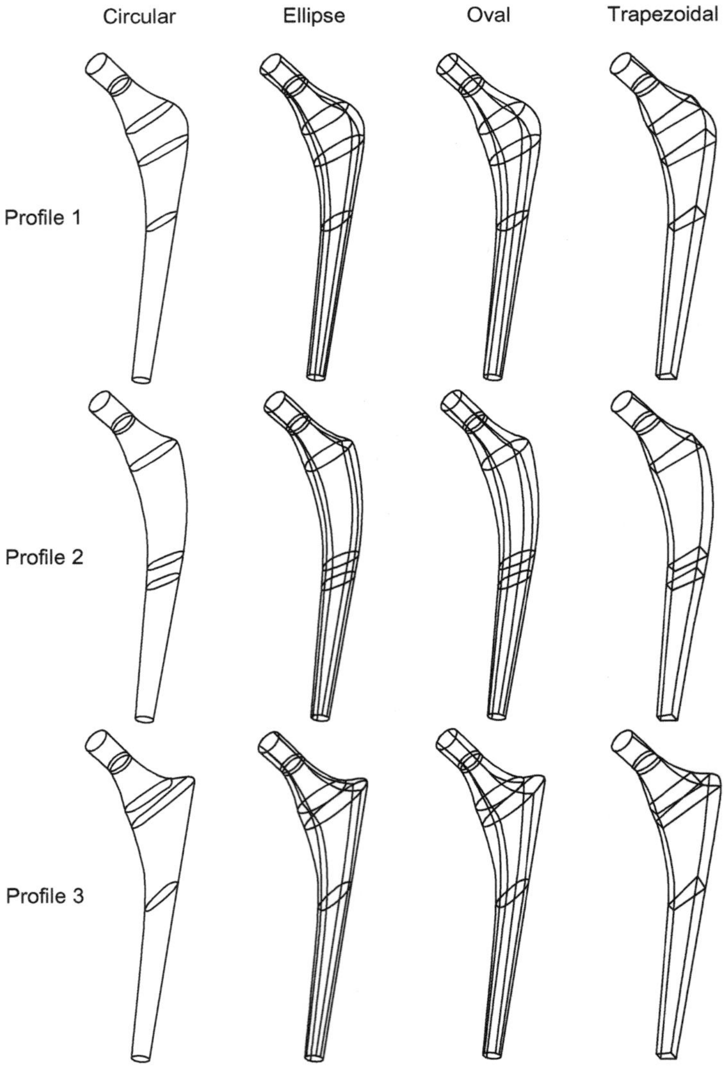

Fig. 2.1 Different stem and femoral neck profiles [11].

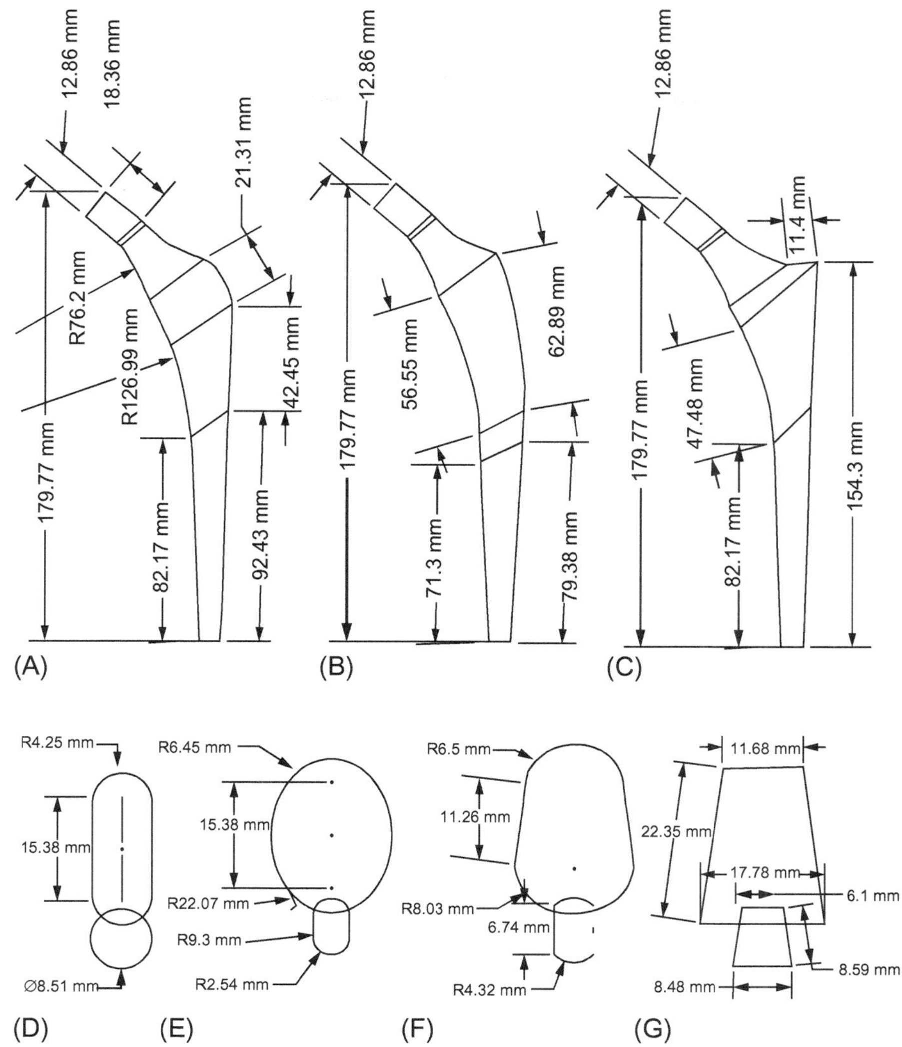

Fig. 2.2 Dimensions of sides and cross section of stem and femoral neck [11].

three different stem and neck profiles with four different cross-sectional shapes, resulting in 12 different designs. This is an efficient design approach; combining many different geometrical ideas and then modeling, analyzing, and comparing them is an effective method for optimizing a design while remaining original. Fig. 2.2 shows the next stage in the design process, adding physical dimensions to the designs. Once inadequate designs have been eliminated, the remaining potential designs are given dimensions and modeled.

As seen in Fig. 2.2, a suitable length and height of the femoral neck are 18.36 and 12.86 mm, respectively. All the cross sections shown have a larger height dimension than width, comprising the entire length of the stem including the neck. The widths of these designs range from 8.50 to 17.78 mm. Another important geometric parameter that needs to be considered in the design of the femoral neck is the angle of inclination. The cross sections shown in Fig. 2.2 are displayed front on for convenience; however, in reality the cross section for a femoral neck will be inclined upwards at an angle. This angle varies widely with the physical requirements of the patient. As there is no set angle, all of the designs are modeled at varying angles to determine an optimum angle for the optimum design. While modeling the contact mechanics of a hip replacement, four cup inclination angles, varying between 35 and 65 degrees in 10-degree increments were considered [12]. Thus it is reasonable to use similar dimensions with the femoral neck design in this investigation. A visual schematic of the inclination angle in relation to the orientation of the load and femoral neck is shown in Fig. 2.3.

Based on the preceding information, the following properties, parameters, and boundary conditions are considered for designing the femoral neck of a hip joint prosthesis:

(a) Ti-6Al-4V ELI alloy will be considered for femoral neck material. The properties of the materials are taken from Table 2.1.
(b) Length 18.36 mm, height 12.86 mm, width 8.50 mm and angles of inclination 35, 45, 55, and 65 degrees will be implemented as geometric dimensions.

2.3 Design procedure

Software package ANSYS Mechanical APDL was used to design and analyze a femoral neck in this investigation. A standard typical set of properties, parameters, and boundary conditions was established for all designs to ensure that each design was compared uniformly, so that consistent properties, parameters, and boundary

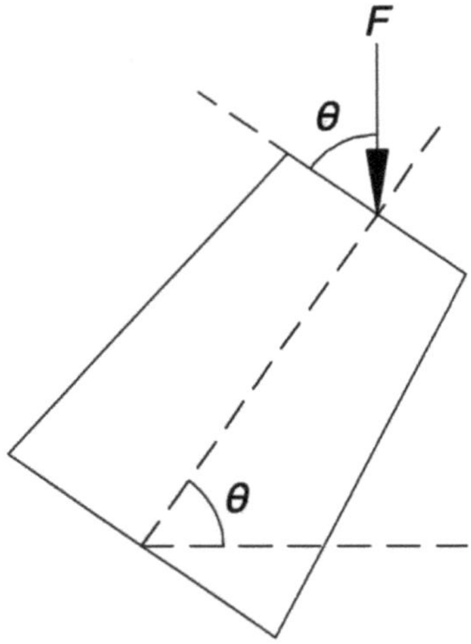

Fig. 2.3 Inclination angle of load application.

conditions were maintained. A typical patient was selected, based on previous studies, as a 70-year-old male weighing 80 kg. It is reasonable to assume that the forces exerted on the femoral neck within a hip replacement lie within the range of 2.5–3.5 times the patient's body weight while walking at a freely selected speed [13]. The forces for the standard patient of 2.5, 3.0, 3.5, and 4.0 times the patient's body weight coincide with the forces 1962, 2354.4, 2746.8, and 3139.2 N.

The variation in femoral neck design was carried out by changing side profiles and cross-sectional geometries and then femoral neck designs were compiled by combining side profiles and cross-sectional geometries. Eleven different femoral neck designs were compiled using different combinations of the side profiles and cross sections and were then modeled in ANSYS. Fig. 2.4 shows the sketches of various side profiles labeled from 1 to 6, while Fig. 2.5 shows the initial preliminary sketches of various cross-sectional designs labeled from a to h. Each of the side profiles was paired with a cross section to compile a design that was then assigned typical dimensions and modeled in ANSYS. The profiles shown in Fig. 2.7 are oriented with an inclination angle of 0 degrees for convenience, but in reality each design was rotated at an angle, as shown in Fig. 2.3. The profiles were rotated from the center of the bottom edge by the specified angle; this means the left side was raised and the right side lowered. The force then acts vertically downwards in the middle of the top face.

The mesh size of each design was evaluated and optimized. Initially all 11 designs were modeled using the automated mesh size in ANSYS, which proved to be inaccurate as the automated mesh differed from one design to the next depending on the complexities of each design. Manual meshes were then created for each design to accurately represent the intended object, until no significant changes in results were noticed with further decrease of the mesh size. The element type SOLID187 is a 10-node element capable of simulating the behavior of a three-dimensional solid body and was used in the modeling of all designs. All designs were modeled as a linear, elastic, isotropic material with an elastic modulus of 101 GPa and Poisson ratio of 0.32, as presented in Table 2.1 [14].

The initial test load and orientation of the product must also be consistent among the designs to give an accurate comparison. The applied force and angle of inclination

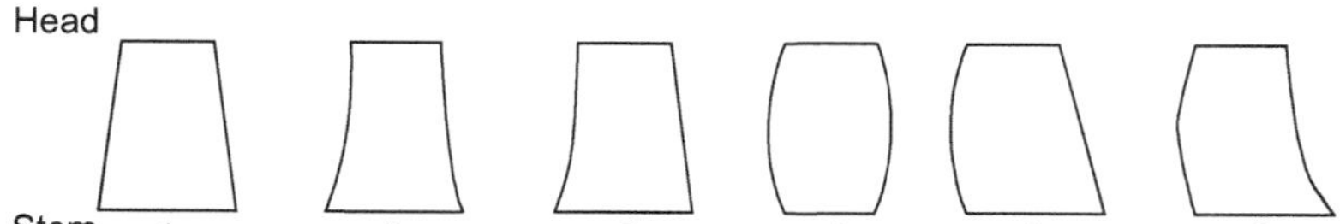

Fig. 2.4 Preliminary side profile schematics.

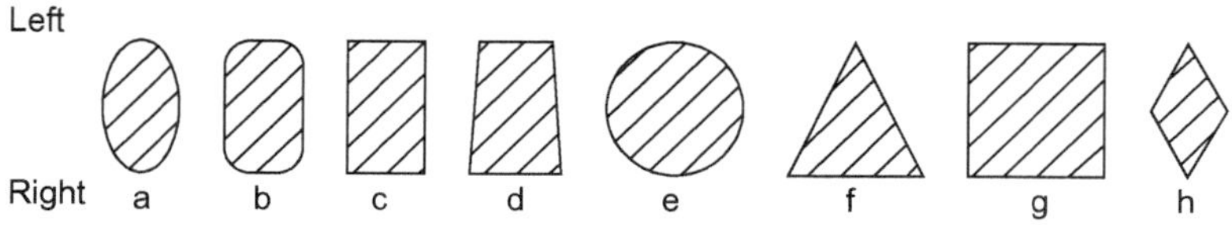

Fig. 2.5 Preliminary cross-section schematics.

started with a load of 1962 N and angle of inclination of 55 degrees, with the remaining loads and angles later being modeled. The base of the neck that connects to the stem was modeled as a fixed surface and the force was applied at the center of the top face that connects to the head at the designated angle. In order to model the force at an angle, the x and y components must be used. For a force of 1962 N acting at a downwards direction at an angle of 55 degrees, the force in the x direction equates to 1125.36 N and the force in the y direction is −1607.18 N. In order to conduct a comparative analysis among designs and determine the most effective design, the minimum stress concentration, minimum deformation and stress distribution of each design was considered. The ability of designs to resist external forces and deformation depends on the intensity of the internal forces, or stresses, of the design [15] and therefore the performance of each design can be estimated by analyzing the stresses and deformation of each design.

The top-performing designs based on the lowest maximum stress, minimum stress, maximum deformation, and stress distribution were summarized. These designs were then modeled with varying load magnitudes (1962, 2354.4, 2746.8, and 3139.2 N) and angles (35, 45, 55, and 65 degrees) to determine the final design. This was necessary because the requirements of a hip replacement are not just in the strength and endurance of the design but also in the versatility. Human beings differ widely from person to person and thus it is important for a hip replacement prosthesis to be applicable with various different loads and angles.

2.4 Results and discussion

A number of femoral neck designs were created by combining the different preliminary side profiles and cross sections illustrated in Figs. 2.4 and 2.5, respectively. There were 48 possible design combinations plus the addition of reversed or mirrored profiles and sections. However, only the most favorable 11 designs were modeled, simulated, and analyzed in this investigation. Fig. 2.6 shows the plotted contours of the Von-Mises stress from the side of the neck distributed throughout the component.

2.4.1 Stress distribution

The stress distribution of each design is represented as a nodal contour plot on the geometry of the design, as depicted in Fig. 2.6. All designs were modeled using the same properties and conditions, and thus the geometry of the designs was the only varying factor. The color range of the contour plot was manually input to ensure that a direct comparison of contours could be made. The color range of the stress contours was set at 0.095–200 MPa for all designs.

The stress distribution pattern was similar across all designs, as shown in the contour plots. All designs have the same load contact location at the center of the top face of the neck, where the maximum stress occurs. This is shown as a red round shape and changes to orange, yellow, green, cyan, light blue, and eventually dark blue as it expands from the center and the stress reduces. The stress distribution pattern on the

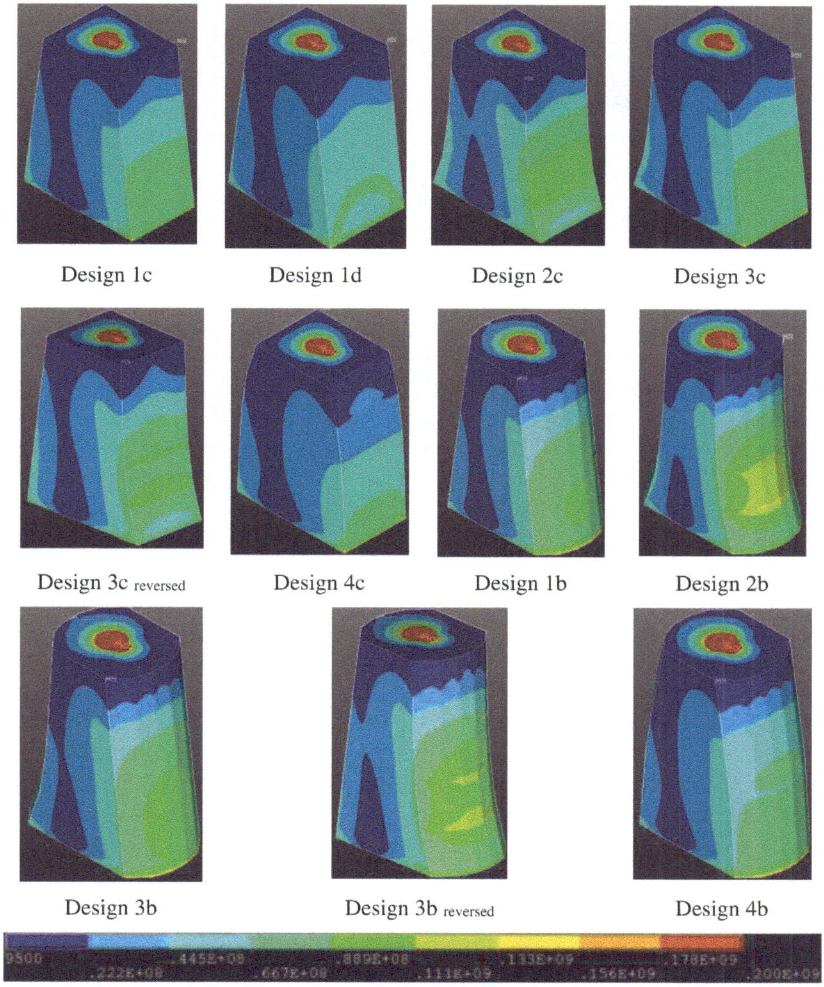

Fig. 2.6 Contour plot of stress generated in different designs due to load application.

front face of the neck is also similar across all designs. The dark blue, or lowest, stress continues down the face from the top of the neck and thins out to a dark blue streak down the middle of the face. On either side of the dark blue strip the stress increases to light blue and cyan towards the left and right faces. This is due to the bending of the neck in reaction to the load. The right face is compressed downwards due to the load acting to the right and down, and the left face is under tension as it is being stretched upwards with the movement of the top of the neck.

The right face of the neck is where the most significant differences exist across the different designs. The dark blue of the top face continues down the right face and changes colors all the way to the bottom of the neck with the increase of stress due to the compression of the right side of the neck. Designs 2b and 3b reversed appear to have the highest amount of stress along the right side, with the stress contour reaching yellow as the neck curves upwards. The exponentially curved neck from the top to the bottom combined with the outward curve of the cross section increases the stress in these designs. Designs 1b, 2c, 3b, and 3c reversed all have the next highest stress in the right side of the neck, all with green stress contours. These four designs have either an exponential curve from the bottom to the top with a straight-edged cross section, or straight sides from bottom to top with a curved-edged cross section. The amount of stress in the remaining five designs all resides in the light green segment of the stress contours. Designs 1c, 3c, and 4b all have light green contours up to roughly half way up the neck, while designs 1d and 4c only have a small section of light green contoured stress, with the majority colored cyan. This indicates that, in terms of visual inspection of the stress contour distribution, designs 1d and 4c are the most effective. The low stress distributions in these designs are due to the trapezium cross section in design 1d and the outward exponentially curved edge from top to bottom in design 4c.

2.4.2 Stress and deformation

In addition to considering distribution of the stresses throughout the products, it is also important to investigate the stresses and deformation values in each design in order to determine the most effective femoral neck design. A product's resistance to deformation and the capacity of the material to resist external forces depend on the intensity of the internal stresses [15]. The ideal femoral neck design is therefore the one that will experience the lowest maximum stress and deformation and the minimum stress during application and that will hence increase the product's performance and life expectancy.

Fig. 2.7 shows a comparison of the maximum stress values obtained for each design. The desired magnitudes of stress are the maximum stresses, as they are the greatest imposition on the strength of the material [15], and hence an optimum design should experience a minimal amount of stress in order to increase the product's life and reduce the probability of failure. Design 1b had by far the highest maximum stress value of all 11 designs, with a maximum stress of 12,800 MPa, while designs 3b

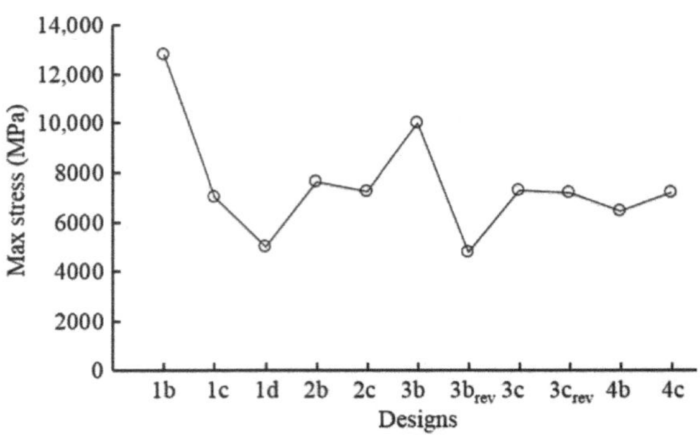

Fig. 2.7 Comparison of maximum stress for all designs.

reversed, 1d, and 4b had the three lowest maximum stress values of 4770, 5020, and 6450 MPa, respectively. Although these individual values may not directly describe the actual maximum stress values, this does suggest that in terms of maximum stress concentration, designs 3b reversed, 1d, and 4b are among the most viable designs.

Fig. 2.8 illustrates a comparison of the minimum stresses for each design. By reducing the overall stress in the design, and hence the minimum stress, the overall quality of the design is improved. Design 1b demonstrates the highest minimum stress concentration in Fig. 2.11 with a minimum stress of 1060 kPa, while designs 1d, 2c, and 3c reversed present the three lowest minimum stresses of 95.66, 153.78, and 157.64 kPa, respectively. Although these individual values may not directly describe the actual minimum stress values, it is clear that designs 1d, 2c, and 3c are the most effective designs in terms of minimum stress concentration.

Fig. 2.9 portrays the maximum deformations of all designs. A reduction in the amount a design deforms during application will reduce the chances of the product failing and hence indicates a design better suited to the problem. Design 1b deflected

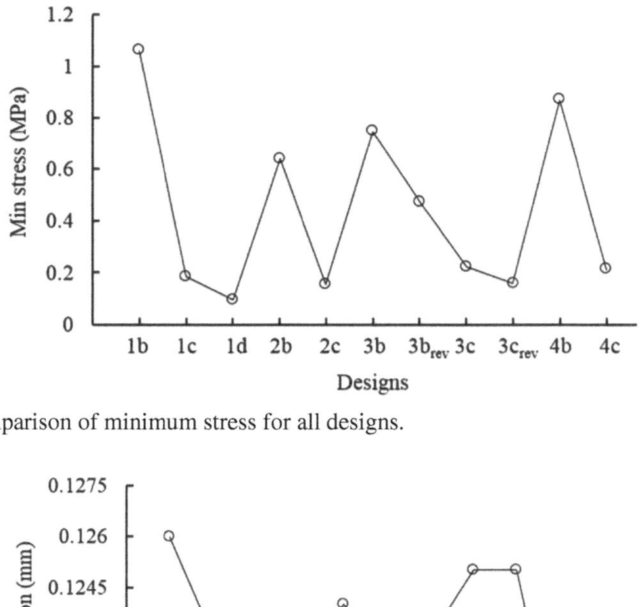

Fig. 2.8 Comparison of minimum stress for all designs.

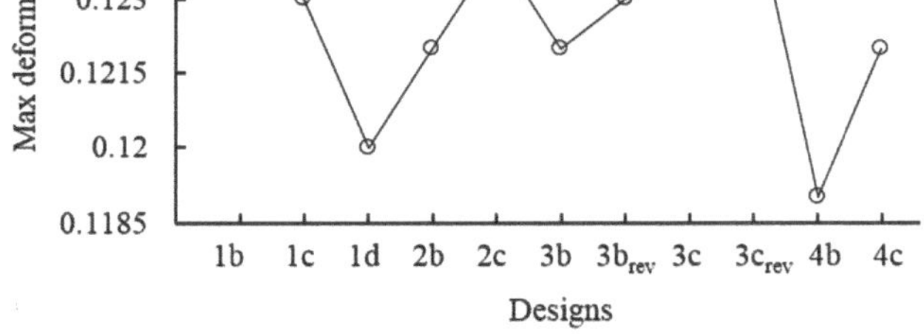

Fig. 2.9 Comparison of maximum deformation for all designs.

the most out of all the designs, with a displacement of 0.126 mm, while designs 4b and 1d moved the least, with deformations of 0.119 and 0.12 mm, respectively. This shows that, although the individual values may not directly describe the actual deformation values, designs 4b and 1d can still be considered the most effective designs in terms of maximum deformation.

2.4.3 Design selection

The top-performing designs were selected as described in the last section on the basis of the lowest maximum and minimum stress values, lowest maximum deformation, and least amount of stress distribution. The resulting top designs from each analysis are summarized in Table 2.2. Each design nominated as one of the top-performing designs in each category, as outlined in the last section, has been allocated one point. A tally of points of the most effective designs in the four different categories is presented in Table 2.2. The two top-performing designs over all four categories are design 1d with four points and design 4b with two points. These two top designs were selected to be modeled with varying load magnitudes and angles to determine the versatility of each design, thus allowing a final design to be declared.

An ideal hip replacement must not only perform effectively by resisting external forces and deformation but also possess the versatility to be applied to various different physical body requirements of patients. The top two designs, design 1d and 4b, were remodeled in ANSYS using the remaining load magnitudes of 2354.4, 2746.8, and 3139.2 N at an inclination angle of 55 degrees and inclination angles of 35, 45, and 65 degrees with a load of 1962 N. This allowed for the prediction of the most versatile design and hence the determination of the final design. Design 1d proved to be the most versatile design by maintaining significantly lower stresses than design 4b with the change in boundary conditions and orientation. Design 1d was therefore the final design of this project, having been demonstrated to be among the highest performing designs with a low amount of internal stresses and deformation, while maintaining the versatility to apply to varying loads and different orientations.

2.4.3.1 Effect of various loads

Figs. 2.10 and 2.11 show the four different contour stress distribution patterns of both designs 1d and 4b, respectively, with loads 1962, 2354.4, 2746.8, and 3139.2 N at an inclination angle of 55 degrees. The figures show that the higher forces increase the

Table 2.2 Summary of highest performing femoral neck designs

Designs	1d	2c	3b Rev	3c	4b	4c
Maximum stress	1	–	1	–	1	–
Minimum stress	1	1	–	1	–	–
Maximum deformation	1	–	–	–	1	–
Stress distribution	1	–	–	–	–	1
Total	4	1	1	1	2	1

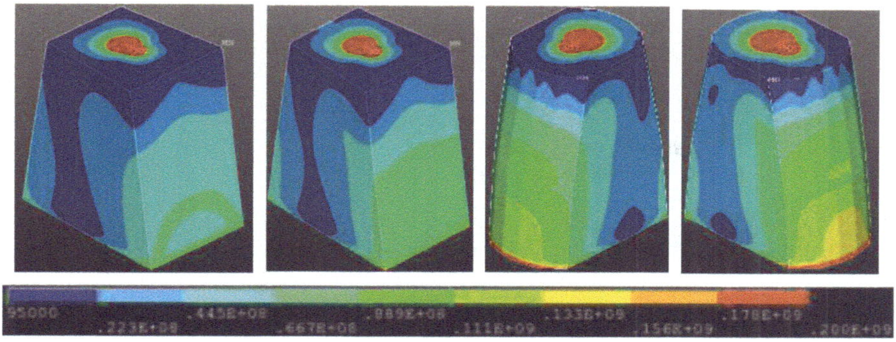

Fig. 2.10 Design 1d with loads 1962, 2354.4, 2746.8, and 3139.2 N at angle 55 degrees.

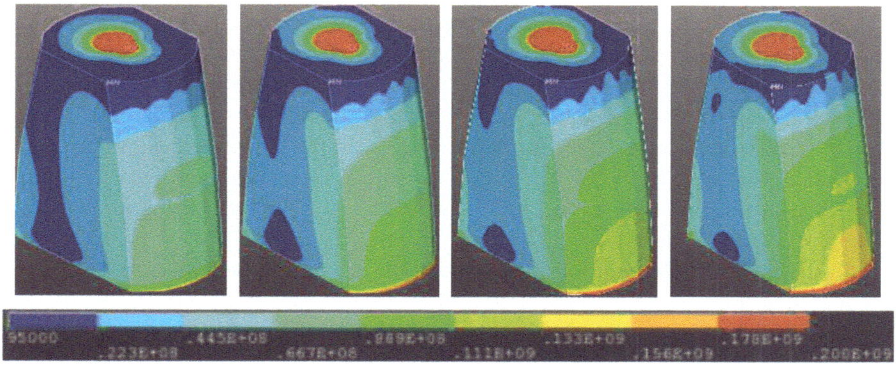

Fig. 2.11 Design 4b with Loads 1962, 2354.4, 2746.8, and 3139.2 N at angle 55 degrees.

stress throughout the design. The stress distribution on the right face of the neck rises noticeably with the increase in load magnitude, changing from predominantly cyan to mostly green in design 1d and from primarily cyan/light green to mainly green/yellow in design 4b. This is to be expected; as the load increases, so will the compression on the right face as the force of the load pushes forward and downward and hence increases the stress on the face. The maximum and minimum stress concentration and the maximum deformation of both designs 1d and 4b also change with the increase of magnitude of the load.

Fig. 2.12 illustrates the variation of the maximum stress of designs 1d and 4b at different loads. Both designs have the same positively linear trend; maximum stress increases with the increase load. The line of the maximum stress for design 1d is however below the line of design 4b. The gradient of the change in maximum stress for design 1d is roughly 2.55 while the gradient for design 4b is approximately 3.27. This shows that not only is the initial maximum stress value for design 1d lower than 4b but the gradient is also, illustrating that as the force is increased further, the maximum stress values of design 1d will continue to be less than those of design 4b.

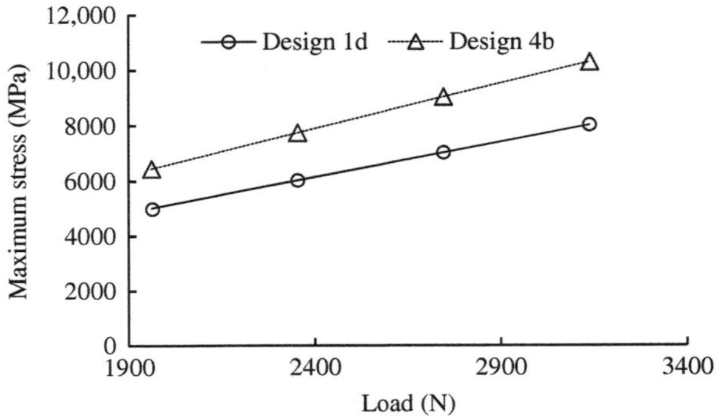

Fig. 2.12 Effect of load on maximum stresses for designs 1d and 4b.

The variation of the minimum stress at different loads for designs 1d and 4b is displayed in Fig. 2.13. Both designs show a positively linear trend and thus the minimum stress of both designs increases with the increase of the load. The magnitude and gradient of both sets of data differ greatly. The initial minimum stress value for design 1d is approximately 0.0957 MPa and it increases with a gradient of roughly 4.87×10^{-5}. The larger initial minimum stress for design 4b is roughly 0.869 MPa and it increases with a much higher gradient of 4.43×10^{-4}. The design 1d has the lowest minimum stress (nine times smaller than design 4b) as well as the minimum stress increase at a much lower rate than design 4b.

Fig. 2.14 presents the relationship between the maximum deformation of designs 1d and 4b and the change in magnitude of the applied load. Both sets of data present very close results with similar magnitudes and gradients for both designs. Design 1d has an initial maximum deformation of 0.12 mm and design 4b has a very comparable initial maximum deformation of 0.119 mm. The gradient of the change in maximum deformation for design 1b is 6.20×10^{-5} and the gradient for design 4b is the slightly lower value of 6.03×10^{-5}. This shows that both designs start at an almost equal initial maximum deformation value, but design 4b, having a slightly lower gradient,

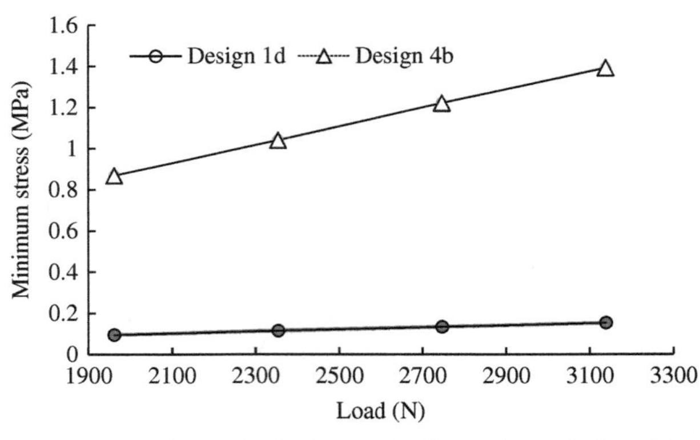

Fig. 2.13 Minimum stress vs change in load magnitude graph for designs 1d and 4b.

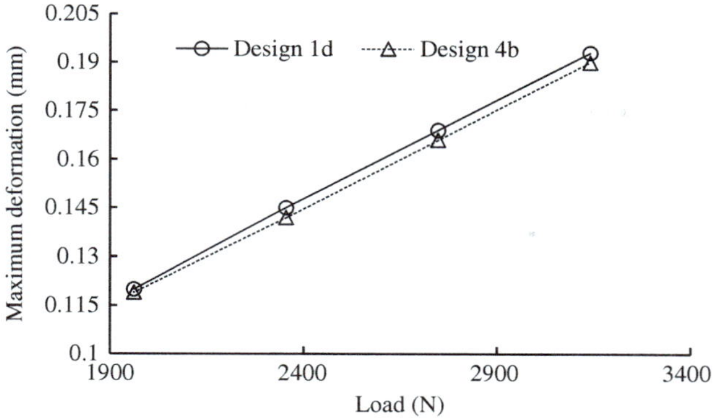

Fig. 2.14 Effect of load on maximum deformation for designs 1d and 4b.

increases at a lower rate and will therefore have a lower deformation as the load magnitude is increased.

2.4.3.2 Effect of various angles of load application

Figs. 2.15 and 2.16 show the four different contour stress distribution patterns of both designs 1d and 4b, respectively, with a load of 1962 N at inclination angles 35, 45, 55, and 65 degrees. Those suggest that the stress distributions of both designs 1d and 4b decrease with the increase of inclination angle. As the inclination angle of the neck is increased, the angle of the downward applied force to the top face of the neck also increases. For the lowest inclination angle, 35 degrees, the right face of design 1d is mostly light green with some green colored contours and for design 4b the right face consists of yellow, green, and light green stress contours. For the highest inclination angle of 65 degrees, the right face of design 1d is now only predominantly cyan and design 4b has a right face mostly colored cyan with some light green contours. This shows that when the force approaches the angle perpendicular to the neutral axis of the neck or *transverse load*, the stress distribution is increased, and when the force

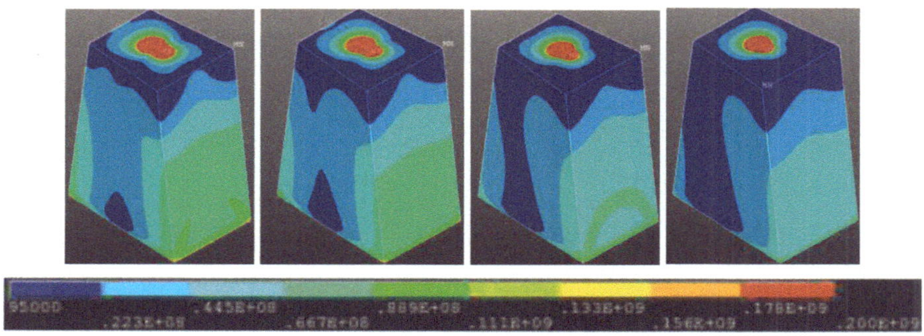

Fig. 2.15 Design 1d at load 1962 N with inclination angles 35, 45, 55, and 65 degrees.

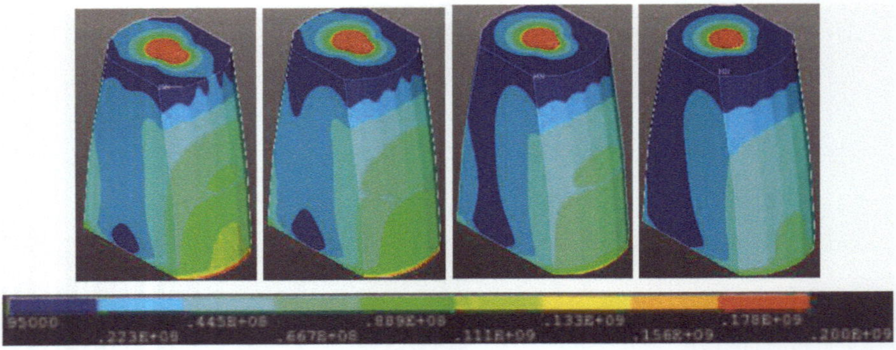

Fig. 2.16 Design 4b at load 1962 N with inclination angles 35, 45, 55, and 65 degrees.

approaches the angle parallel to the neutral axis of the neck or *axial load*, the stress distribution is decreased. This is an expected outcome; the neck of the prosthesis can withstand axial loads more effectively and is more subject to failure when exposed to transverse loads. This is because transverse loads induce internal axial forces, shearing forces, and bending moments [15] in the structure, increasing the internal stresses throughout.

Fig. 2.17 illustrates the variation of maximum stress of designs 1d and 4b at different inclination angles. Both designs show a decrease in maximum stress with the increase of the inclination angle. These results are expected; as the inclination angle of the design and hence the angle of the applied load is increased, it moves towards the axial position of the neck and hence the stress is likely to reduce. The initial maximum stress of design 1d is 6550 MPa at 35 degrees and is lower than that of design 4b of 8760 MPa at 35 degrees. The overall gradient of the data for design 1d is −81.67 compared to a steeper gradient of −124.67 for design 4b. This suggests that if the inclination angle is further increased, the maximum stress of design 4b will be equal to that of 1d.

The variation of the minimum stress at different inclination angles for designs 1d and 4b are displayed in Fig. 2.18. Both designs display an overall decrease in minimum

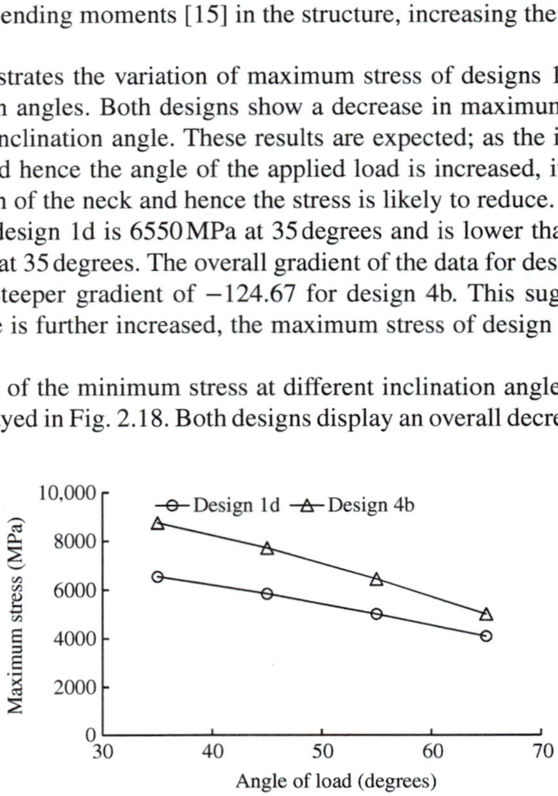

Fig. 2.17 Effect of inclination angle on the maximum stress for designs 1d and 4b.

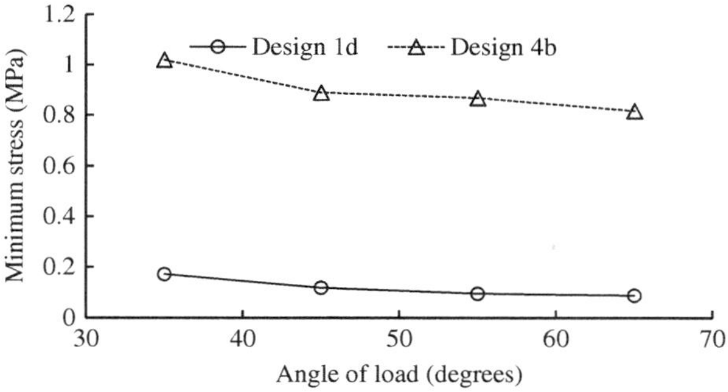

Fig. 2.18 Effect of inclination angle on minimum stress for designs 1d and 4b.

stress concentration with the increase of inclination angle. Design 1d shows a negative trend with an initial minimum stress value of 0.173 MPa for 35 degrees. Design 4b has the much higher initial minimum stress value of 1.02 MPa for 35 degrees. The overall data for design 4b also decreases. The difference in magnitude between designs 1d and 4b is significant with the entire graph of design 1d being entirely below that of design 4b. These results are expected; the stresses in the femoral neck are minimized as the angle of the applied load approaches 90 degrees or parallel to the length of the neck.

Fig. 2.19 presents variation of maximum deformation of designs 1d and 4b with the change in inclination angle. Both datasets demonstrate a negative linear trend and thus as the angle of the inclination is increased, the maximum deformation of the design is decreased. Both graphs have very similar trends with a small difference in magnitude; design 1d has an initial maximum deformation of 0.129 mm and design 4b has the slightly lower initial deformation of 0.128 mm. The trends are almost the same, with design 1d breaking the trend with a shallower gradient at 65 degrees than design 4b. The decrease in maximum deformation with the increase in angle of inclination is an expected outcome for both designs. As the inclination angle is increased, so is the angle between the applied load and the surface of the top face of the neck. By increasing this angle, the y force component is increased and the x force component is decreased. The femoral neck experiences more deformation by forces in the

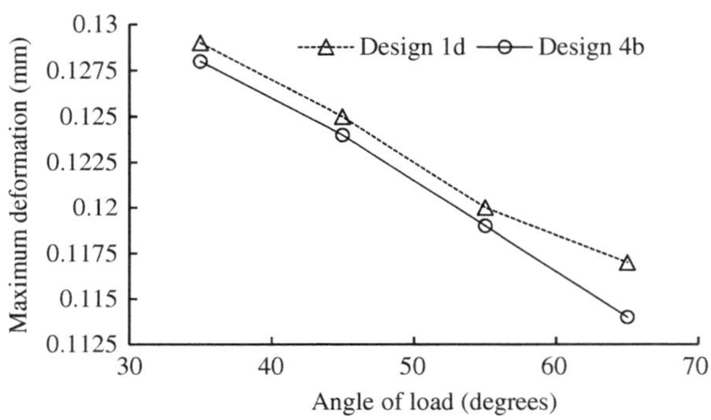

Fig. 2.19 Effect of inclination angle on maximum deformation for designs 1d and 4b.

x direction than the y direction, due to its vertical shape and the induced bending and shear forces, and hence, as the inclination angle is increased, the less the applied load deflects the design.

2.5 Conclusions

Eleven different femoral neck designs were modeled and analyzed using ANSYS Mechanical APDL. After a comparative analysis of the maximum and minimum stress concentration, maximum deformation, and stress distribution across all designs, the top two designs were selected. Designs 1d and 4b demonstrate a low amount of stress concentration and deformation and show a decreased stress distribution pattern, indicating that these designs are the strongest and the most durable of the original 11 designs.

The versatility of the top two designs was then tested by modeling different load magnitudes and inclination angles to demonstrate varying physical requirements of a hip replacement design. For both the varying load magnitude and the varying inclination angle data, design 1d proved to have a significant decrease in maximum and minimum stress concentrations, as well as stress distribution, in comparison to design 4b. However, design 4b did portray a slight decrease in maximum deformation in comparison to design 1d for both varying load magnitude and inclination angle. The significance of the lower stress concentration and distribution of design 1d far outweighs the slight decrease in maximum deformation of design 4b, thus indicating that design 1d is more likely to adapt to various boundary conditions and is therefore the more versatile of the two. Design 1d is the most suitable design in effectiveness and versatility and is therefore the final design for this investigation.

References

[1] S.R. Knight, R. Aujla, S.P. Biswas, Total hip arthroplast—over 100 years of operative history, Orthop. Rev. 3 (2) (2011) 72–74.

[2] G.D. Baura, Total hip prostheses, in: Medical Device Technologies, ScienceDirect, 2012, pp. 381–404. (Chapter 18).

[3] M.R. Baumgaertner, T.F. Higgins, Fractures of the neck of femur, in: R.W. Bucholz, J.D. Heckman, C.M. Court-Brown (Eds.), Rockwood and Green's fractures in adults, sixth ed., Lippincott Williams & Wilkins, Philadelphia, 2005, pp. 1579–1634.

[4] A.D. Woolf, B. Plefger, Barden of major musculoskeletal condition, Bull. World Health Organ. 81 (9) (2003) 646–656.

[5] A. Freitas, R.A. Maciel, R.D.A. Lima, D.R.D.M. Souto, M.D.A. Ferrer, Mechanical analysis of femoral neck fracture fixation with dynamic condylar screw in synthetic bone, Acta Ortop. Bras. 22 (5) (2014) 264–268.

[6] L.S. Estrada, D.A. Volgas, J.P. Stannard, J.E. Alonso, Fixation failure in femoral neck fractures, Clin. Orthop. Relat. Res. 399 (2002) 110–118.

[7] S.A. Asgari, A. Hamouda, S. Mansor, H. Singh, E. Mahdi, R. Wirza, Finite element modeling of a generic stemless hip implant design in comparison with conventional hip implants, Finite Elem. Anal. Des. 40 (2004) 2027–2047.

[8] A. Pramanik, L.C. Zhang, Residual stresses in silicon-on-sapphire thin film systems, Int. J. Solids Struct. 48 (9) (2011) 1290–1300.

[9] M. Niinomi, Mechanical properties of biomedical titanium alloys, Mater. Sci. Eng. 243 (1) (1998) 231–236.

[10] M. Semlitsch, Titanium alloys for hip joint replacements, Clin. Mater. 2 (1987) 1–13.

[11] A.A. Oshkour, N.A.A. Osman, M. Bayat, R. Afshar, F. Berto, Three-dimensional finite element analyses of functionally graded femoral prostheses with different geometrical configurations, Mater. Des. 56 (2014) 998–1008.

[12] X. Hua, J. Li, L. Wang, Z. Jin, R. Wilcox, J. Fisher, Contact mechanics of modular metal-on-polyethylene total hip replacement under adverse edge conditions, J. Biomech. 47 (13) (2014) 3303–3309.

[13] R.A. Brand, D.R. Pedersen, D.T. Davy, G.M. Kotzar, K.G. Heiple, V.M. Goldberg, Comparison of hip force calculations and measurements in the same patient, J. Arthroplast. 9 (1) (1994) 45–51.

[14] Titanium Alloy Guide, RMI Titanium Company, http://www.rtiintl.com/Titanium/RTI-Titanium-Alloy-Guide.pdf, 2000 (Accessed May 6, 2015).

[15] E.P. Popov, S. Nagarajan, Z.A. Lu, Mechanics of Materials, second ed., Prentice-Hall, Upper Saddle River, NJ, 1976.

Mechanical properties of the optic nerve head

3

Junfei Tong, Linxia Gu
Department of Mechanical and Materials Engineering, University of Nebraska-Lincoln, Lincoln, NE, United States

3.1 Introduction

Glaucoma is an eye disease characterized by nonreversible visual field loss, and the elevated intraocular pressure (IOP) is well-acknowledged to be responsible for its initiation and development [1]. The optic nerve head (ONH) is the location where retinal axon bundles (prelaminar neural tissue) pass through and form the optic nerve, as shown in Fig. 3.1. Lamina cribrosa (LC), a porous connective tissue supporting and nourishing the optic nerve, has been proposed as the primary site of axonal injury in the ONH [2]. The IOP induced stress and strain have been proposed to play a significant role in the damage and remodeling of the ONH tissue [1]. Therefore, the biomechanics of LC induced by IOP has been widely studied through both experiments and computational simulations. Specifically, Hua et al. performed a sensitivity analysis on the biomechanics of ONH using a generic computational model, and found the sclera material property was one of the highest determinant factors for ONH biomechanics [3].

With increasing attention focused on glaucoma in the past decades, more glaucomatous cases without elevated IOP, also referred to as normal tension glaucoma, were reported, while some individuals with elevated IOP failed to develop glaucoma [3]. It was then believed that the intracranial pressure (ICP) in the subarachnoid space might also play a role in the development of glaucoma. It was reported by Berdahl et al. that ICP was significantly higher in nonglaucoma patients as compared to glaucoma patients [4]. In another study, it was reported that normal tension glaucoma patients were found to have significantly lower ICP, while the ocular hypertension group had significantly higher ICP [3]. Those facts indicate that ICP might play an important role in the mechanics of ONH. The translaminar pressure difference, i.e., the difference between IOP and ICP, instead of either IOP or ICP alone, has been correlated with the visual field related to the neuroretinal rim area [5].

The material properties of the ONH tissues determining the biomechanics of the ONH are of great importance to the understanding of the glaucoma initiation and development. Downs et al. reported that the peripapillary sclera in early-glaucoma eyes has a significantly higher equilibrium modulus compared to normal eyes [6]. After the exposure to moderately elevated IOP, significantly stiffer sclera was also observed by Girard et al. [7]. The influence of material properties on the biomechanics of ONH was also investigated by computational methods [8–10]. The material properties of sclera, LC, and dura mater were found to impact the ONH biomechanics the most.

Mechanical Behavior of Biomaterials. https://doi.org/10.1016/B978-0-08-102174-3.00003-6

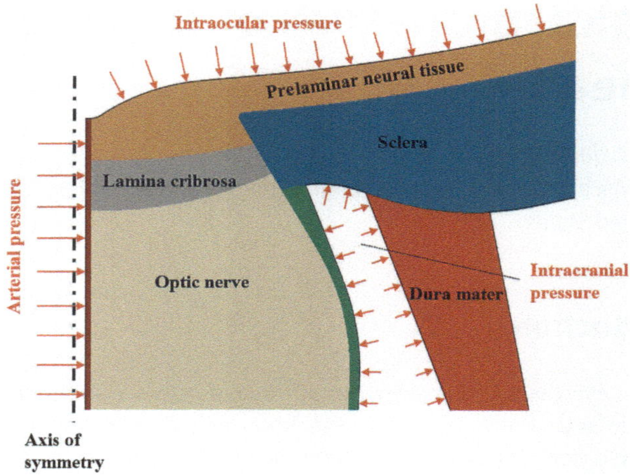

Fig. 3.1 The geometry and loading condition of the ONH.

In this chapter, we reviewed the mechanical properties of the ONH tissues that have the highest significance, specifically, the sclera, LC and dura mater.

3.2 Material testing methods

3.2.1 Sclera

The sclera is the outer layer of the eye, which consists of collagen and glycosaminoglycans. The material property of the sclera is highly associated with the microstructural arrangement of collagen. At a higher collagen fiber density and along the fiber direction, the sclera exhibited greater stiffness [11]. For the human posterior sclera, collagen fiber architecture was highly anisotropic and inhomogeneous [12]. The elevation of IOP may induce regional differences of fiber alignment in the posterior sclera between glaucomatous and nonglaucomatous eyes. A comprehensive understanding of sclera biomechanical behavior is important for better clinical strategies.

The mechanical behaviors of the sclera were the most studied among all the components of the ONH as summarized in Table 3.1. Significant variations were observed. This could be attributed to different testing protocols including the testing methods, sample preparation, age, and interspecimen variations.

3.2.1.1 Uniaxial testing

Uniaxial testing is the simplest and most straightforward method in determining material properties. A hyperelastic behavior of sclera was generally observed, as it becomes stiffer at a larger strain or stress. For most of the unfixed sclera, the uniaxial test was conducted within 7 days of harvest without much degradation [23]. During the test, the test samples were normally placed in 37°C saline to maintain the hydration,

Table 3.1 Material characterizations of the sclera

	Number of samples	Age range (mean)	Test method	Treatment	Tangent modulus	Ref.
Human sclera	36	52–96	Uniaxial test	Medium Eusol-C at 4°C tested within 5–7 days	At σ=0.2 MPa, ~14.4 MPa for anterior specimen, ~13.7 MPa for equatorial specimen, ~11 MPa for posterior specimen	[13]
	45	51–84	Uniaxial test	Medium Eusol-C at 4°C tested within 5 days	At σ=0.05 MPa, ~10 MPa; at σ=1 MPa, ~60 MPa	[14]
	10	66–89 (83)	Uniaxial test	At 4°C for 30 days	At ε=1%, ~2.6 MPa	[15]
	5	45–63	Uniaxial test		At ε=8%, 22–55 MPa	[16]
	10	50–59 (55)	Biaxial test	Normal saline tested within 45 h	Stiffness product: ~2.9 MPa along latitudinal (circumferential) direction, ~2.8 MPa along (meridian) longitudinal direction	[17]
	24	30–74	Uniaxial test	Tested in 37°C saline with 1 mm/s	Heel modulus: ~4.4 MPa along vertical direction, ~4.5 MPa along horizontal direction	[18]
	48		Uniaxial test	DMEM+0.5 M glyceraldehyde for 6 days DMEM+0.5 M methylglyoxal for 6 days DMEM+0.1% collagenase for 1 day	Control: 28.9 MPa, glyceraldehyde: 65.5 MPa Control: 36 MPa, methylglyoxal: 159.6 MPa Control: 28.5 MPa, collagenase: 16.5 MPa	[19]

Continued

Table 3.1 Continued

	Number of samples	Age range (mean)	Test method	Treatment	Tangent modulus	Ref.
Porcine sclera	10	6–8 months	Uniaxial test	At 4°C tested within 48 h of enucleation	At $\varepsilon = 1\%$, ~3 MPa; at peak load, ~17 MPa	[20]
	10	6–8 months	Inflation test	At 4°C tested within 48 h of enucleation	At $\varepsilon = 1\%$, ~6 MPa; at peak load, ~15 MPa	[20]
	13		Uniaxial test	At 4°C for 4 days	At $\varepsilon = 1\%$, ~0.65 MPa	[15]
	50		Uniaxial test		At $\varepsilon = 8\%$, 4–53 MPa	[16]
	8		Biaxial test	Stored in phosphate buffered saline (PBS) at 4°C tested within 72 h	Stiffness product: ~0.48 MPa along circumferential direction, ~0.34 MPa along meridian direction	[21]
	180	5–7 month	Uniaxial test	Tested within 18 h	Heel modulus: ~5.7 MPa along vertical direction, ~4.4 MPa along horizontal direction	[22]
	80		Uniaxial test	DMEM+0.5 M glyceraldehyde for 6 days DMEM+0.5 M methylglyoxal for 6 days DMEM+0.1% collagenase for 1 day DMEM+0.1% glutaraldehyde for 30 min	Control: 29.3 MPa Glyceraldehyde: 54.7 MPa Methylglyoxal: 61.5 MPa Glutaraldehyde: 56.8 MPa Collagenase: 3.4 MPa	[19]

as lower temperature saline or air circumference was found to significantly increase the porcine sclera stiffness [22]. The adopted strain rate also influenced the test result. Elsheikh et al. found that the tangent modulus of the sclera increased by >30% as the strain rate rose from 8% to 200% per min [13].

Biomaterial properties generally exhibited interspecimen and intraspecimen variations. Elsheikh et al. found that the tangent stiffness of the human sclera increases from posterior regions towards the limbus using a uniaxial tensile test [13]. Geraghty et al. also observed that the posterior sclera was ~24% stiffer than that of the equator, and sclera in the latter was 54% stiffer than that in the anterior region [14]. The sclera is isotropic at the equator and anisotropic near the ONH, with collagen fiber along the circumferential direction [24]. On the contrary, human [18] and porcine posterior sclera [22] were also reported to exhibit similar mechanical behavior along the vertical and horizontal directions. The possible reason for the controversy could be the different test methods.

A prevalence of glaucoma was found in the aging population. Friedman et al. reported that prevalence of open-angle glaucoma in the United States was <1% in those aged under 55 years, but increased to 3.8%–7.7% in those aged over 75 [25]. Similarly, Rudnicka et al. summarized 46 clinical trials related to glaucoma and found that the prevalence of glaucoma increased with age for all racial groups, including White, African American, and Asian [26]. This motivated further studies on the role of age on the biomechanical behavior of the sclera.

It was found that the stiffness of the human sclera increased with age in all regions using uniaxial tests [14]. Grytz and Fazio also reported similar findings using inflation tests and computational models [27, 28]. More details on inflation testing are given in Section 3.2.1.3. The age-induced stiffening in the sclera could be explained by the stiffening of the collagen caused by advanced glycation end-products (AGEs), which were accelerated in the diabetic group due to the high level of glucose [29]. Schultz et al. reported that the tangent modulus of the human sclera at 1% strain was 75% higher than that of the porcine sclera. This might be attributed to higher nonenzymatic cross-link (glycation) density in the human sclera [15]. Spoerl et al. discovered that the stiffness of the human sclera increased from 28.9 to 65.5 MPa after incubation with glyceraldehyde for 6 days. The same applied to the porcine sclera [19].

3.2.1.2 Biaxial testing

Tissues are generally subjected to multiaxial loading conditions in vivo. Biaxial loading was designed to couple the material performance along two perpendicular directions, which could better mimic the in vivo mechanical environments. In these tests, a planar square specimen was generally immersed in saline maintained at body temperature and stretched along two orthogonal directions. During the testing, the tethering effect at the outer edges of the specimen, which is similar to the edge effect at uniaxial grips, was excluded from the postprocessing [30]. It is worth noting that the test results could be influenced by the natural curvature variations of the specimen as well as the collagen alignment in relation to the biaxial loading directions. In addition, the strain uniformity in the central region of the specimen was highly sensitive to grip details [31].

The constitutive model could be described by the Fung's strain energy density function [17, 21] as follows:

$$W = \frac{C}{2}\left(e^{\left(C_1E_{11}^2 + C_2E_{22}^2 + 2*C_3E_{11}E_{22}\right)} - 1\right) \tag{3.1}$$

$$S_{ij} = \frac{\partial W}{\partial E_{ij}} \tag{3.2}$$

where W is the strain energy, E_{ij} are the Green strains, S_{ij} are the second Piola-Kirchoff stresses, and C_i are the material constants. For the biaxial testing set-up,

$$S_{11} = C\left(C_1E_{11} + C_3E_{22}\right)e^{\left(C_1E_{11}^2 + C_2E_{22}^2 + 2*C_3E_{11}E_{22}\right)} \tag{3.3}$$

$$S_{22} = C\left(C_2E_{22} + C_3E_{11}\right)e^{\left(C_1E_{11}^2 + C_2E_{22}^2 + 2*C_3E_{11}E_{22}\right)} \tag{3.4}$$

$$\frac{S_{11}}{S_{22}} = \frac{C_1E_{11} + C_3E_{22}}{C_2E_{22} + C_3E_{11}} \tag{3.5}$$

Regression analysis, such as least square method, was used to obtain the material constants from the testing results. For the case of equal-biaxial tests, i.e., $E_{11} = E_{22}$, the ratio between S_{11} and S_{22} would reduce to $\frac{C_1 + C_3}{C_2 + C_3}$, which indicates the level of anisotropy.

Only two groups performed biaxial testing of the sclera, as summarized in Table 3.1. Eilaghi et al. measured 10 human sclera specimens ranging from 50 to 59 years old, and found no significant difference in sclera between latitudinal direction and longitudinal direction [17]. It is worth noting that the stiffness of sclera obtained from the biaxial test was less than that from the uniaxial test. Perez et al. performed a biaxial testing on eight porcine sclera specimens, where the strain was measured through high-resolution ultrasound speckle tracking instead of using a traditional displacement sensor [21]. The circumferential stiffness was found to be significantly higher than that of the meridian, and the stiffness along the two orthogonal directions was positively correlated ($R^2 = 0.8$). This could be related to the previous observation that the preferred alignment for collagen fibers was along the circumferential direction [24]. The discrepancy of the sclera anisotropy between the aforementioned datasets might be due to the different sclera regions. Specifically, the sclera specimen used by Eilaghi et al. was 3–9 mm from the ONH center, while in Perez's study, it was 2 mm from the ONH center.

3.2.1.3 Inflation testing

The inflation test of the posterior eyeball is a more physiologically sound way to quantify the biomechanical behavior of the sclera. A digital image correlation or laser speckle interferometry was commonly utilized to measure the three-dimensional displacement and further the strain distributions of the sclera [27, 28, 32, 33]. The material properties of the sclera could then be derived using an inverse finite element analysis.

Lari et al. investigated the measurement differences between an inflation test and a uniaxial test of a porcine sclera [20]. It was found out that the peak strain induced by the uniaxial test was more than doubled that of the inflation test with equivalent loading. Correspondingly, the sclera tangential modulus obtained from the inflation test was 6 MPa, which is twice that of the uniaxial test at a 1% strain. Fazio et al. characterized the strain variations in different sclera regions using an inflation test [32]. As the IOP increased from 5 to 45 mmHg, the peripapillary sclera had a significantly higher tensile strain (1.28%) than the midperipheral sclera (0.98%). The high strain in the peripapillary sclera might protect the LC from damage. As the posterior or peripapillary sclera became stiffer with age [13, 14], the reduction in strain would be expected. The reduced strain was associated with a large deformation in the LC and a higher risk of glaucoma [28].

Coudrillier et al. evaluated the age effect on the collagen fiber structure and material property of seven human sclera using an inflation test [34]. The three-dimensional displacement was calculated with digital image correlation and the collagen fiber structure was mapped with wide-angle X-ray scattering. Based on the displacement map and collagen fiber orientation, an inverse finite element analysis was developed to minimize the difference between experimental displacement and model prediction. This was used to estimate the material properties of collagen and the interfiber matrix. It was found that age was strongly associated with a lower degree of fiber alignment, and therefore a reduced anisotropy. And then near-isotropy in peripapillary sclera could induce larger scleral canal expansion and larger tensile and shear strain in the LC [33]. This indicated that the loss of fiber alignment with aging might induce more damage on the retinal ganglion cells leading to higher risk of glaucoma.

3.2.2 *Lamina cribrosa*

The LC is a porous connective tissue consisting of collagen and elastin, through which the retinal ganglion cell axon bundles pass and form the optic nerve. Its biomechanical performance was highly dependent on the stiffness of the fibers rather than on the microstructural alignment [24, 35]. Specifically, doubled collagen stiffness reduced the strain in neural tissue by 38%, while the fiber anisotropy had minimal influence on the neural tissue [35]. Considering the small size of the LC, its material testing is challenging. Only a couple of studies focused on the experimental tests of the LC properties [19, 36]. In addition, a few more studies have evaluated the biomechanical behaviors of the LC noninvasively using a laser scanning microscope or the inverse finite element method as shown in Table 3.2 [37–40].

Similar to the sclera, the LC was also found to become stiffer with aging, due to the AGE-induced stiffening of the collagen [29, 39]. The increased rigidity of the LC has been associated with glaucoma progression [41]. To further characterize the influence of AGEs on the LC biomechanical behavior, Spoerl et al. evaluated the stiffness changes of both human and porcine LC that were disposed of in aldehyde sugars, such as glyceraldehyde, methylglyoxal, and glutaraldehyde (porcine only) [19]. It was observed that the stiffness of the LC increased by ~50% to 200% after the treatment (Table 3.2). This indicated that nonenzymatic glycation in aging

Table 3.2 Material characterizations of the LC

	Number of samples	Age	Test method	Treatment	Tangent modulus	Ref.
Human LC	48		Uniaxial test	DMEM+0.5 M glyceraldehyde for 6 days DMEM+0.5 M methylglyoxal for 6 days DMEM+0.1% collagenase for 1 day	Control: 14.7 MPa, glyceraldehyde: 24.2 MPa Control: 11.8 MPa, methylglyoxal: 296.2 MPa Control: 15.6 MPa, collagenase: 9.8 MPa	[19]
	1	57	Model fitting	Fitted to artificial displacement	0.19 MPa	[37]
			Model fitting	Fitted to experiment done by Yan et al.	0.14 MPa at 5 mmHg 0.38 MPa at 50 mmHg	[38]
	6	64–92	AFM	Tested in PBS	Control: 17.2 kPa Pseudoexfoliation: 10.1 kPa	[36]
	10	7–86	Inflation	Inflated with saline	Not quantitatively evaluated	[39]
	8	26–93	Inflation	Inflated with water	Not quantitatively evaluated	[40]
Porcine LC	80		Uniaxial test	DMEM+0.5 M glyceraldehyde for 6 days DMEM+0.5 M methylglyoxal for 6 days DMEM+0.1% collagenase for 1 day DMEM+0.1% glutaraldehyde for 30 min	Control: 17.1 MPa Glyceraldehyde: 22.4 MPa Methylglyoxal: 66.4 MPa Collagenase: 7.6 MPa Glutaraldehyde: 28.8 MPa	[19]

could increase the stiffness of the collagen, and thus the tissue as a whole. The LC stiffening with aging agrees with the experimental observation of a lower strain in aged LC [40].

The monitored geometrical changes in the LC could be used to estimate its material properties using the inverse finite element method. Edwards et al. [38] developed a mathematical model to evaluate the biomechanical behavior of the LC, which was validated by the experiment [42]. The Young's modulus of the LC increased from 0.14 MPa at IOP = 5 mmHg to 0.38 MPa at IOP = 50 mmHg. This indicated that the LC is a hyperelastic material. The increased stiffness of the LC with increasing IOP was also observed by Julie et al. using a laser scanning confocal microscope [39]. The lower stiffness at small load was attributed to the collagen fiber uncramping. The fiber stretch-induced the increased stiffness at larger loading. Such a process could minimize the damage to neural tissue [43].

The reverse-fitting of the LC stiffness was also performed by Zhang et al. by using the virtual field method. The stiffness of the LC was evaluated as 0.19 MPa based on the ONH displacement [37]. This magnitude was much smaller than the one tested by Spoerl et al. [19]. Braunsmann et al. evaluated the material properties of human LC and sclera by using atomic force microscopy (AFM) [36]. The stiffness of the LC was one order of magnitude smaller than that evaluated through inverse finite element modeling or tensile test [19, 37, 38]. The stiffness of the sclera was two orders of magnitude lower than that tested by tensile test [16, 19, 20]. This was also reported in the material testing of the trabecular meshwork (TM). The stiffness of the porcine TM was 1.38 kPa using the AFM, while it was 2490 kPa using a tensile test [44, 45]. Such discrepancy of the testing results may be attributed to the different test methods. For example, the AFM focused on the locational properties, and its testing result is dependent on the test protocol (e.g., tip size, tissue thickness, working force) [46].

3.2.3 Dura mater

Dura mater is a fibrous tissue protecting the brain, spinal cord, and the optic nerve. It is composed of collagen and elastin and is flexible to accommodate the stretch and deformation caused by cerebrospinal fluid (CSF) pressure changes [47]. The dura fiber anisotropy varies with its location. The human lumbar dura fiber has a preferred alignment along the longitudinal direction, while human cranial dura is close to isotropic [48–51]. The anisotropy in the lumbar dura may be attributed to longitudinal stretch when the spine bends anteriorly, while for cranial dura, the stretch is more uniform in both directions and thus it shows isotropic behavior [50].

The mechanical properties of the dura mater attracted lots of attention in the 1970s because of its demonstrated performance as a heart valve replacement [48, 52–54]. Recently, dura mater was investigated towards better understanding of the tissue damage caused by traumatic injury, spinal cord injury, and subdural and epidural hematomas [55–57]. The material characterizations of the dura mater are summarized in Table 3.3. Significant variations between different groups as well as different species are exhibited.

Glycerol was a widely used agent for preserving the dura mater before surgery, including heart valve replacement. Van Noort et al. evaluated the influence of glycerol

Table 3.3 Summary of material properties of the dura mater

Species	Number of samples	Age range	Location	Test method	Treatment	Tangent modulus	Ref.
Human	6	39–86 (58.5)	Lumbar	Uniaxial	Tested within 120h frozen at −4°C test strain: 20%–60%	Longitudinal stiffness: 60–100MPa circumferential stiffness: <10MPa	[50]
	16	20–77		Uniaxial	Preserved in 98% glycerol test strain: 16%–20%	21.3–48MPa	[54]
	13	17–72	Brain	Uniaxial	Preserved in 98% glycerol test strain: not available	Fresh: 61.5MPa, glycerol: 45.2MPa	[48]
Porcine	28		Cervix	Uniaxial	Preserved in 0.9% normal saline at 4°C for <24h; Test strain: <50%	not quantitatively evaluated	[57]
	15		Optic nerve	Uniaxial	Testing strain <14%, Yeoh model	C1 = 0.17MPa, C2 = 4.21MPa, C3 = −4.97MPa	[58]
Ovine	12		Spine	Biaxial	Testing strain <12%, anisotropic	Circumferential direction: 1.13MPa Longitudinal direction: 0.32MPa	[55]

Bovine	38	16–24 month	Spine	Uniaxial	Testing strain <60%, Ogden model	Circumferential direction: 0–6 MPa (instant shear modulus) Longitudinal direction: 0–0.04 MPa (instant shear modulus)	[59]
	9		Spine	Relaxation	Testing strain: 2%–46%, quasilinear theory	Circumferential direction: 92–380 MPa Longitudinal direction: 0.2–3.4 MPa	[60]
			Lumbar	Uniaxial	Tested within 96 h frozen at −4°C Test strain: 80%–120%	Longitudinal stiffness: 25–80 MPa	[50]
Rat		77 ± 5 days	Spine and brain	Uniaxial and relaxation	Testing strain <50%, Ogden model	Spinal dura: 1.20 MPa (instant shear modulus) Cranial dura: 0.42 MPa (instant shear modulus)	[61]

or saline on the mechanical behavior of the human dura mater [53, 54]. The elastic modulus of the human dura mater ranged from 21.3 to 48 MPa, regardless of preservation in glycerol for 12 or 30 days. Similar results were also observed for the saline. McGarvey et al. observed that the glycerol resulted in a reduced stiffness from 61.5 to 45.2 MPa, while such a change was not significant. This indicated that glycerol and saline could be used as a storage or test medium without altering the material properties of the dura mater. Moreover, the tensile strength of the dura mater was found decreased with age ($R=0.79$). This implied a deterioration process with aging for dura mater [53].

Most studies investigated the mechanical behavior of the dura mater in the spine or lumbar (Table 3.3), while only one study focused on the brain dura mater [48]. It was found that cranial dura mater is completely isotropic. For dura mater at the lumbar or spine, the collagen fiber was found to be more aligned along the longitudinal direction in human or other species [50, 59, 61]. The dura mater in the circumferential direction had a low crimped-to-straight length ratio (1.21), compared to that of the longitudinal direction (1.66) [59]. Because of that, the dura mater exhibits stiffer behavior at low strain in the circumferential direction [55, 57, 59, 60]. As the fiber was fully stretched, significant stiffer dura mater was observed along the longitudinal direction [50, 59]. An interesting finding in porcine cervical dura mater was that it was close to isotropic on the dorsal side but tends to be stiffer in the circumferential direction on the ventral side. The discrepancy may be attributed to its location, as the dura mater was from vertebra C2, which was mainly responsible for rotation, where the ventral side contributes more in maintaining the vertebral stability and thus might be stiffer along the circumferential direction. It worth noting that mechanical properties from biaxial tests in ovine were 1–2 orders of magnitude lower than those from a uniaxial test in other species. The exact reason for such phenomenon is yet unknown, it may be attributed to the natural variability of dura mater between different species.

Dura mater was not sensitive to the strain rate [48, 59–61]. Wilcox et al. observed no significant changes in the stiffness of bovine dura mater through stress relaxation tests at strain rates ranging from 2% to 46% per min. McGarvey et al. evaluated the human cranial dura through a stress relaxation test with strain rate ranging from 5% to 500% per min, where they found no significant difference in the stress–strain curve for fresh or preserved dura mater [48]. A similar result was also observed by Persson et al. in bovine spinal cord dura mater using a uniaxial test with strain rate as 0.01, 0.1, and $1\,s^{-1}$; no significant difference was found in the Ogden model parameters under different strain rate [59]. As mentioned in the sclera section, strain rate was found to significantly influence the sclera mechanical properties [13]; the strain rate influence discrepancy between dura mater and sclera may be attributed to the large natural variability of the biomaterial itself.

The mechanical behavior of the dura mater was important in computational modeling to understand the mechanism of glaucoma as well as visual impairment and intracranial pressure (VIIP) syndrome [9, 10]. While the current material test for dura mater is mainly focused on cranial and spinal dura mater, only few studies had tested the mechanical properties of optic nerve dura mater [58]. As aforementioned, the location has a significant influence on the dura mater microstructure alignment in response to loading conditions. While the microstructure alignment of the optic nerve dura mater was poorly studied. Based on the circular loading from CSF, the loading condition of dura mater is similar to that of artery. Since the inner layer of artery has a preferred alignment along the circumferential direction [62], we thus speculate the optic nerve dura mater may also have a similar alignment. The understanding of the dura mater microstructure alignment is important as it may explain the regional difference of the dura mater material properties. Maiko et al. evaluated the dura mater of rat in both spine and brain; they found that spinal dura mater (1.2 MPa) was significantly stiffer than cranial dura mater (0.42 MPa) [61]. In addition, an accurate characterization of the optic nerve dura mater is of great important for the numerical modeling of the ocular diseases [58].

3.3 Summary

The sclera was the most studied component in the ONH. The testing methods included tensile testing, indentation testing, relaxation testing, and inflation testing. The specimens span different scales, ages, locations, and species. A hyperelastic anisotropic sclera was generally observed. Even though one aspect of mechanical behaviors is generally adopted in existing studies, a full spectrum of the material characterization might be helpful for better clinical strategies [54].

The difficulty in sample preparation of the LC was reflected in the few numbers of publications. More datasets are desirable for the LC. It is known that LC in the ONH subjected to both IOP and ICP that have diurnal fluctuations. However, the dynamic mechanical properties (e.g., viscoelasticity) of the LC has never been studied. Despite Albon et al. and Midgett et al. observing a stiffer LC with increasing age [39, 40], no quantitative data is available for the material properties of the LC in response to aging and regional variations.

The mechanical behavior of the dura mater in the lumbar or spine has been widely studied, but few studies have investigated that in the optic nerve dura mater. Since the dura mater in different regions has different microstructure as well as material properties, it would be highly important to investigate both the microstructure and material properties in optic nerve dura mater. The detailed characterization of these components would provide better understanding of the mechanism of glaucoma, which could be transferred to other eye diseases, such as VIIP syndrome and traumatic ocular injury.

References

[1] C.F. Burgoyne, et al., The optic nerve head as a biomechanical structure: a new paradigm for understanding the role of IOP-related stress and strain in the pathophysiology of glaucomatous optic nerve head damage, Prog. Retin. Eye Res. 24 (1) (2005) 39–73.

[2] H.A. Quigley, et al., Optic nerve damage in human glaucoma: II. The site of injury and susceptibility to damage, Arch. Ophthalmol. 99 (4) (1981) 635–649.

[3] J.P. Berdahl, et al., Intracranial pressure in primary open angle glaucoma, normal tension glaucoma, and ocular hypertension: a case–control study, Invest. Ophthalmol. Vis. Sci. 49 (12) (2008) 5412–5418.

[4] J.P. Berdahl, R.R. Allingham, D.H. Johnson, Cerebrospinal fluid pressure is decreased in primary open-angle glaucoma, Ophthalmology 115 (5) (2008) 763–768.

[5] R. Ren, et al., Trans-lamina cribrosa pressure difference correlated with neuroretinal rim area in glaucoma, Graefes Arch. Clin. Exp. Ophthalmol. 249 (7) (2011) 1057–1063.

[6] J.C. Downs, et al., Viscoelastic material properties of the peripapillary sclera in normal and early-glaucoma monkey eyes, Invest. Ophthalmol. Vis. Sci. 46 (2) (2005) 540–546.

[7] M.J. Girard, et al., Biomechanical changes in the sclera of monkey eyes exposed to chronic IOP elevations, Invest. Ophthalmol. Vis. Sci. 52 (8) (2011) 5656–5669.

[8] I.A. Sigal, J.G. Flanagan, C.R. Ethier, Factors influencing optic nerve head biomechanics, Invest. Ophthalmol. Vis. Sci. 46 (11) (2005) 4189–4199.

[9] A.J. Feola, et al., Finite element modeling of factors influencing optic nerve head deformation due to intracranial pressure, Invest. Ophthalmol. Vis. Sci. 57 (4) (2016) 1901–1911.

[10] Y. Hua, et al., Intracranial pressure influences the behavior of the optic nerve head, J. Biomech. Eng. 139 (3) (2017) 031003.

[11] S. Lin, L. Gu, Influence of crosslink density and stiffness on mechanical properties of type I collagen gel, Materials 8 (2) (2015) 551–560.

[12] J.K. Pijanka, et al., Quantitative mapping of collagen fiber orientation in non-glaucoma and glaucoma posterior human sclerae, Invest. Ophthalmol. Vis. Sci. 53 (9) (2012) 5258–5270.

[13] A. Elsheikh, et al., Regional variation in the biomechanical properties of the human sclera, Exp. Eye Res. 90 (5) (2010) 624–633.

[14] B. Geraghty, et al., Age-related variations in the biomechanical properties of human sclera, J. Mech. Behav. Biomed. Mater. 16 (2012) 181–191.

[15] D.S. Schultz, et al., Structural factors that mediate scleral stiffness, Invest. Ophthalmol. Vis. Sci. 49 (10) (2008) 4232–4236.

[16] G. Wollensak, E. Spoerl, Collagen crosslinking of human and porcine sclera, J. Cataract Refract. Surg. 30 (3) (2004) 689–695.

[17] A. Eilaghi, et al., Biaxial mechanical testing of human sclera, J. Biomech. 43 (9) (2010) 1696–1701.

[18] K. Chen, et al., Elastic properties of human posterior eye, J. Biomed. Mater. Res. A 102 (6) (2014) 2001–2007.

[19] E. Spoerl, A.G. Boehm, L.E. Pillunat, The influence of various substances on the biomechanical behavior of lamina cribrosa and peripapillary sclera, Invest. Ophthalmol. Vis. Sci. 46 (4) (2005) 1286–1290.

[20] D.R. Lari, et al., Scleral mechanics: Comparing whole globe inflation and uniaxial testing, Exp. Eye Res. 94 (1) (2012) 128–135.

[21] B.C. Perez, et al., Biaxial mechanical testing of posterior sclera using high-resolution ultrasound speckle tracking for strain measurements, J. Biomech. 47 (5) (2014) 1151–1156.

[22] K. Chen, A.P. Rowley, J.D. Weiland, Elastic properties of porcine ocular posterior soft tissues, J. Biomed. Mater. Res. A 93 (2) (2010) 634–645.

[23] L. Fontana, et al., Frequency of positive donor rim cultures after penetrating keratoplasty using hypothermic and organ-cultured donor corneas, Cornea 26 (5) (2007) 552–556.

[24] L. Zhang, et al., Collagen microstructural factors influencing optic nerve head biomechanics, Invest. Ophthalmol. Vis. Sci. 56 (3) (2015) 2031–2042.

[25] D.S. Friedman, et al., Prevalence of open-angle glaucoma among adults in the United States, Arch Ophthalmol. 122 (4) (2004) 532–538.
[26] A.R. Rudnicka, et al., Variations in primary open-angle glaucoma prevalence by age, gender, and race: a Bayesian meta-analysis, Invest. Ophthalmol. Vis. Sci. 47 (10) (2006) 4254–4261.
[27] R. Grytz, et al., Age-and race-related differences in human scleral material properties, Invest. Ophthalmol. Vis. Sci. 55 (12) (2014) 8163–8172.
[28] M.A. Fazio, et al., Age-related changes in human peripapillary scleral strain, Biomech. Model. Mechanobiol. 13 (3) (2014) 551–563.
[29] A.J. Bailey, R.G. Paul, L. Knott, Mechanisms of maturation and ageing of collagen, Mech. Ageing Dev. 106 (1) (1998) 1–56.
[30] M.S. Sacks, W. Sun, Multiaxial mechanical behavior of biological materials, Annu. Rev. Biomed. Eng. 5 (1) (2003) 251–284.
[31] A. Eilaghi, et al., Strain uniformity in biaxial specimens is highly sensitive to attachment details, J. Biomech. Eng. 131 (9) (2009) 091003.
[32] M.A. Fazio, et al., Regional variations in mechanical strain in the posterior human sclera, Invest. Ophthalmol. Vis. Sci. 53 (9) (2012) 5326–5333.
[33] B. Coudrillier, et al., Scleral anisotropy and its effects on the mechanical response of the optic nerve head, Biomech. Model. Mechanobiol. 12 (5) (2013) 941–963.
[34] B. Coudrillier, et al., Collagen structure and mechanical properties of the human sclera: analysis for the effects of age, J. Biomech. Eng. 137 (4) (2015) 041006.
[35] A. Voorhees, N.-J. Jan, I. Sigal, Effects of collagen microstructure and material properties on the deformation of the neural tissues of the lamina cribrosa, Acta Biomater. 58 (2017) 278–290.
[36] C. Braunsmann, et al., Evaluation of lamina cribrosa and peripapillary sclera stiffness in pseudoexfoliation and normal eyes by atomic force microscopy, Invest. Ophthalmol. Vis. Sci. 53 (6) (2012) 2960–2967.
[37] L. Zhang, et al., Verification of a virtual fields method to extract the mechanical properties of human optic nerve head tissues in vivo, Biomech. Model. Mechanobiol. 16 (3) (2017) 871–887.
[38] M.E. Edwards, T.A. Good, Use of a mathematical model to estimate stress and strain during elevated pressure induced lamina cribrosa deformation, Curr. Eye Res. 23 (3) (2001) 215–225.
[39] J. Albon, et al., Age related compliance of the lamina cribrosa in human eyes, Br. J. Ophthalmol. 84 (3) (2000) 318–323.
[40] D.E. Midgett, et al., The pressure-induced deformation response of the human lamina cribrosa: analysis of regional variations, Acta Biomater. 53 (2017) 123–139.
[41] R.C. Zeimer, Y. Ogura, The relation between glaucomatous damage and optic nerve head mechanical compliance, Arch. Ophthalmol. 107 (8) (1989) 1232–1234.
[42] D. Yan, et al., Deformation of the lamina cribrosa by elevated intraocular pressure, Br. J. Ophthalmol. 78 (8) (1994) 643–648.
[43] N.-J. Jan, et al., Microstructural crimp of the lamina cribrosa and peripapillary sclera collagen fibers, Invest. Ophthalmol. Vis. Sci. 58 (9) (2017) 3378–3388.
[44] F. Yuan, L.J. Camras, P. Gonzalez, Trabecular meshwork stiffness in ex vivo perfused porcine eyes, Invest. Ophthalmol. Vis. Sci. 52 (14) (2011) 6693.
[45] L.J. Camras, et al., Differential effects of trabecular meshwork stiffness on outflow facility in normal human and porcine eyes, Invest. Ophthalmol. Vis. Sci. 53 (9) (2012) 5242–5250.

[46] D. Sicard, L.E. Fredenburgh, D.J. Tschumperlin, Measured pulmonary arterial tissue stiffness is highly sensitive to afm indenter dimensions, J. Mech. Behav. Biomed. Mater. 74 (2017) 118–127.

[47] F. Vandenabeele, J. Creemers, I. Lambrichts, Ultrastructure of the human spinal arachnoid mater and dura mater, J. Anat. 189 (Pt 2) (1996) 417.

[48] K.A. McGarvey, J.M. Lee, D.R. Boughner, Mechanical suitability of glycerol-preserved human dura mater for construction of prosthetic cardiac valves, Biomaterials 5 (2) (1984) 109–117.

[49] D.J. Patin, et al., Anatomic and biomechanical properties of human lumbar dura mater, Anesth. Analg. 76 (3) (1993) 535–540.

[50] M. Runza, et al., Lumbar dura mater biomechanics: experimental characterization and scanning electron microscopy observations, Anesth. Analg. 88 (6) (1999) 1317–1321.

[51] M.C.J. Hamann, M.S. Sacks, T.I. Malinin, Quantification of the collagen fibre architecture of human cranial dura mater, J. Anat. 192 (1) (1998) 99–106.

[52] K. Nakiri, et al., Dura mater valve for cardiac prostheses, ASAIO J. 21 (1) (1975) 573–580.

[53] R. Van Noort, et al., The mechanical properties of human dura mater and the effects of storage media, Clin. Phys. Physiol. Meas. 2 (3) (1981) 197.

[54] R. Van Noort, et al., A study of the uniaxial mechanical properties of human dura mater preserved in glycerol, Biomaterials 2 (1) (1981) 41–45.

[55] S.S. Shetye, M.M. Deault, C.M. Puttlitz, Biaxial response of ovine spinal cord dura mater, J. Mech. Behav. Biomed. Mater. 34 (2014) 146–153.

[56] C. Cha, J. Oh, An optofluidic mechanical system for elasticity measurement of thin biological tissues, Biotechnol. Lett. 35 (5) (2013) 825–830.

[57] E. Mazgajczyk, et al., Mechanical properties of cervical dura mater, Acta Bioeng. Biomech. 14 (1) (2012) 51–58.

[58] X. Wang, et al., Finite element analysis predicts large optic nerve head strains during horizontal eye movements, Invest. Ophthalmol. Vis. Sci. 57 (6) (2016) 2452–2462.

[59] C. Persson, et al., Poisson's ratio and strain rate dependency of the constitutive behavior of spinal dura mater, Ann. Biomed. Eng. 38 (3) (2010) 975–983.

[60] R.K. Wilcox, et al., Mathematical model for the viscoelastic properties of dura mater, J. Orthop. Sci. 8 (3) (2003) 432–434.

[61] J.T. Maikos, R.A. Elias, D.I. Shreiber, Mechanical properties of dura mater from the rat brain and spinal cord, J. Neurotrauma 25 (1) (2008) 38–51.

[62] L. Azinfar, et al., High resolution imaging of the fibrous microstructure in bovine common carotid artery using optical polarization tractography, J. Biophotonics 10 (2) (2017) 231–241.

Metallic biomaterials—A review

4

*Divya Zindani**, *Kaushik Kumar*†, *J. Paulo Davim*‡
*Department of Mechanical Engineering, NIT Silchar, Silchar, India, †Department of Mechanical Engineering, BIT Mesra, Ranchi, India, ‡Department of Mechanical Engineering, University of Aveiro, Aveiro, Portugal

4.1 Introduction

The aging population has grown exponentially of recent, particularly in developed countries. This has led to drastic increases in the usage of implants, to meet the needs and desires of the aging population to maintain life quality and related activities. Further, there have been urgent requirements to meet the challenges associated with vascular therapy, trauma, cardiology, dental, orthopedics, wound care, etc. To address the different associated challenges, one of the major focal points of the medical community has been on high-performance implantable biomaterials. The market for biomaterials has grown from $95 billion in 2011 to $135 billion in the year 2017 (http://www.marketsandmarkets.com/PressReleases/bio-implants.asp). A rapid surge in the diversified and improved functionalities of biomaterials has taken place with the advancements in natural, synthetic, and hybrid materials. Together with the growth in the different types of biomaterials, there have been rapid advances in processing techniques of biomaterials, making them suitable to be used as implantable devices [1–4]. The presence of diversified biomaterials has allowed for selection of suitable materials that meet particular objectives, as for instance using metals as electrodes for artificial organs with high electroconductivity, materials for permanent replacement having the characteristics of being chemically inert, or materials with biodegradability properties for the cases in which regenerating of lost tissues or restoration of function is possible [5, 6].

In recent years there has been an intense focus on choosing biomaterials with multifunctionality. For instance, a temporary material for a scaffold may not only lend itself to providing physical support to promote tissue regeneration, but may also provide for other critical biological factors such as fibroblast, morphogenetic protein-2, the growth factors for endothelial growth, etc. One example exhibiting such multifunctionality is the magnesium implant [7–10], which has the required tensile strength, fracture resistance, etc. The kinetics of biodegradation can be controlled by the alloying metals as well as through coatings and emerging mechanical processing for the materials. For instance, surface modifications such as coatings with polymer thin films and with bioactive ceramics can provide multifunctionality to the bioinert metals such as Co and Ti based alloys. Such bioinert materials that are based on Co or Ti are mainly used for load-bearing applications where properties such as corrosion resistance can provide enhanced mechanical stability and reliability and minimized local toxicity to the host, both at the local and systemic levels [11, 12]. The Co and Ti based bioinert materials

Mechanical Behavior of Biomaterials. https://doi.org/10.1016/B978-0-08-102174-3.00004-8

possess excellent tensile strength, fatigue stress, and fracture toughness [13, 14]. The multifunctionality bioinert materials have found applications in orthopedics as plates, screws, and artificial joints, as braces and dental implants for orthodontics, as components for neurosurgical devices such as staples, artificial hearts, wires, and stents. In comparison to Co-Cr alloys, the Ti based bioinert materials are the most preferred alternatives because of favorable combinations of corrosion resistance, biocompatibility, elastic modulus, strength, and relatively low weight and density [15].

However, a number of challenges are associated with the applications of diversified implantable materials. One of the challenges is that of controlling the kinetics of biodegradation for some metals, such as Mg or Fe, whereby an early degradation can have an adverse effect on the restoration of the tissue due to loss in mechanical strength. A risk of developing systemic hypersensitivity reactions due to the long-term presence of Co-Cr or Ti alloys in the body may also exist. Osteopenia may result due to stress shielding resulting from modulus differences between these metals and the natural bone tissue [11]. One of the major risks associated with these implants is that of inflammation and infection [16–18]. Such associated risks may undermine the efficiency of the implants and also can lead to loss of tissue in and around the implant [19]. The septic or aseptic loosening in the case of load-bearing implants may lead to incorrect transfer of force, such as that of biting force to the dental implants. The incorrect transfer of load to the implant may lead to its failure once the load goes beyond its fatigue strength [20]. In 20% of the cases of patients with 7–10 years of implant placement, complications such as periimplantitis have been reported [21]. The aforementioned complications associated with the implants can lead to significant health issues in the patients and require additional costs to minimize the complications [22, 23].

A considerable effort has been made to do away with these shortfalls, such as the development of different surface and bulk modification strategies such as acid and plasma etching, laser ablation, ion implantation, etc. The methods for modifying the implants are being researched heavily by the scientific community [24]. The development of 3D printing technology has led to fabrication of biomaterials that have complex geometries and are customized according to the customer requirements [25, 26]. The 3D printing technology can also lead to replication of nano- and microscale features within the material. However, the main challenge of 3D printing technology is to enable the usage of nanoparticles and their melting, which are crucial for the complete assembly. Further, it is difficult to assess the impact of post- and preprocessing done on the implants using 3D printing technology.

The present work explores metallic biomaterials and the key strategies that have been adopted for surface and bulk modifications.

4.2 Permanent metallic bioimplants

The commonly used bioinert metals for angioplasty, fracture fixation, and remodeling of bone are Co-Cr alloys, Ti alloys, and surgical stainless steel [14]. These materials have numerous advantages such as excellent mechanical properties and therefore long-term stability in the highly reactive environment. However, the highly reactive environment in which they function may lead to their degradation, which may in turn result in the release of undesired metallic ions. The released metallic ions cause inflammatory reactions such as osteolysis and may even lead to metal hypersensitivity [27]. The osteolysis undermines the fixation and interferes with the load transfer, resulting in failure of the implant and complications related to surgeries.

4.2.1 Stainless steel-based bioimplants

Pure metals were considered as suitable material candidates for the fabrication of implants before the introduction of surgical stainless steel. The pure metals had lower mechanical strength as well as low corrosion resistance [28]. The limitations associated with the pure metals were overcome by the introduction of 18/8 stainless steel [29] which had relatively better corrosion resistance resulting in fewer complications after the surgery. The high percentage of chromium content, i.e., 12%, is the main reason behind the higher corrosion resistance of stainless steel. Together with the chromium content, stainless steel also possesses molybdenum and nickel. The formation of carbides of chromium takes place during high temperature treatment, which is taken care of by the limited carbon percentage (http://www.worldstainless.org/Files/ISSF/Education/Module_03_Corrosion_Resistance_of_Stainless_Steels.pdf).

Surgical stainless steel can be classified into Ni-free stainless steel or conventional stainless steel. Although the presence of nickel increases corrosion resistance, it also reduces the biocompatibility [30]. Therefore, to restrict the percentage of nickel, stainless steel is alloyed with nitrogen. The conventional stainless steel is used for load-bearing applications as well as for stents; however, the Ni-free stainless steels are used only for stents. The conventional stainless steel is less biocompatible and corrosion resistant than the titanium-based biomaterials, but the cost of titanium-based biomaterials is relatively higher than that of the stainless steel biomaterials. Also, stents function in a unique oxygen-rich environment where the risk of corrosion is minimized, and therefore the stents from stainless steel are more affordable and appropriate [31]. The carbon content in surgical stainless steel is $<0.03\%$, which is about one-fifth of that present in other metallic biomaterials [32]. The better mechanical properties and ductility of stainless steel allow its usage for bone fracture treatments, where it is used in the form of nails, screws, etc. Fracture plates made from stainless steel provide for temporary support for tissue regeneration.

316L stainless steel is used for fabrication of low-cost copies of actual implants, which are used by surgeons for determination of the actual size of the implant during joint replacement surgery. The traditional approach entailed the use of actual implants, which required repeated cycles of decontamination and sterilization. The mechanical strength of the implant would be undermined because of the fatigue and corrosion induced due to the sterilization and decontamination process [33]. Frame proposed a different approach using laser sintering in fabrication of disposable low-cost copies of implants with 316L stainless steel.

SLS/SLM has been used in fabrication of dental implants with stainless steel. Liquid phase sintering is employed to melt the binding polymer with the aid of a laser beam, which is then used to bind the metal particles. The residual polymer is then removed by heating the scaffold [34]. The implant of required density is then produced by further infiltration and sintering of bronze.

4.2.2 Titanium and titanium alloy-based bioimplants

Medical-grade titanium is a much better alternative for high loading rate applications because of its higher strength-to-weight ratio than stainless steel. One of the critical factors for the degree of distress to which the adjacent bones will be subjected is the weight of the alloy [35]. The strength of Ti alloys may be further enhanced with thermal aging, quenching, and annealing [36].

Alloying of Ti is another technique to enhance the mechanical strength of Ti material. For instance, alloying with 5.5–6.5 wt% aluminum and 3.5–4.5 wt% vanadium results in Ti-6Al-4V, which produces a Ti material that has better strength properties than the unalloyed Ti metal [37]. The hardness of Ti increases by 32% with the addition of Al, the other properties remaining intact (http://www.totalmateria.com/article126.htm). The strength as well as corrosion resistance of Ti increases with the addition of niobium [38, 39]. However, in cases where bending is of utmost importance for the implant function, the reduced ductility of Ti arising due to alloying leads to its unsuitability for such applications.

Ti lends itself for restoration of anatomically complex areas and its suitability for such applications is further enhanced with the involvement of 3D printing technology. The repair of bifrontal skull defects [40] and reconstruction of the orbital and maxillary floor [41] have been made possible with the 3D-printed titanium implants. The 3D-printed implants offer low operation time, superior aesthetics, long-term stability, and elimination of facial dysfunction in patients in comparison to the conventional system of screws and plates [42]. However, the 3D-printed titanium implants also show susceptibility to infection, as revealed in reconstruction of zygomatico-orbital defects [43].

A customized 3D-printed titanium prosthesis has also been produced for limb salvage surgery. The titanium prosthesis was used to replace the bones lost to scapular Ewing's sarcoma [44]. The implants from Ti-6Al-4V powder were fabricated using an electron beam system. The usage of an electron beam melting system overcomes the challenges arising because of the liquid Ti showing chemical affinity to atmospheric gases and also because of the low ductility of the Ti metal [45]. In order to bring the excessively high moduli closer to that of cortical bone, porosity was introduced into the structure. This was done to minimize stress shielding rather than promoting vascularization or tissue regeneration.

Titanium-based implants have also been produced using direct metal laser sintering (DMLS). The implants produced using DMLS have been used to support bar-retained maxillary overdentures, the survival rates being 97% and the biological complications at 7% [46]. The survival rates for unsplinted DMLS-produced Ti implants supporting ball attachment for mandibular overdentures were also reported to be similar [46, 47].

The integration of the gradients for porosity and Young's modulus for 3D Ti-6Al-4V dental implants is one of the main advantages of the selective laser sintering technique [48]. Sacrificial wax templates can be used for fabrication of Ti scaffolds with improved anisotropic properties, such as compressive strength and porosity [49]. The durability of the scaffolds is affected significantly by the potentiality to reduce the discrepancies in architecture of the scaffolds [50].

The fatigue resistance of the fabricated dental implants is comparatively more than that of the highly porous titanium implants, which have been obtained using the traditional coating and spraying techniques. The fatigue resistance is 30% lower for the implants produced using traditional methods than those fabricated using the 3D printing technologies. The 3D printing techniques such as SLS also render simultaneous control over the micro- and nanostructures within the geometry as a whole. The other techniques such as solid-state foaming by expansion of the pores filled with argon, sintering of Ti fibers, etc., don't offer simultaneous control [51]. Implants with osseoconductive properties as well as biocompatibility characteristics are produced using the 3D printing techniques [52]. Improved bioresponse and osseoconductivity were shown by beagles implanted with Ti-6Al-4V implants as compared to acid-etched or alumina-blasted implants [53].

4.2.3 Cobalt-based bioimplants

The resistance of Co-based implants is higher in comparison to the Ti alloy-based implants. The enhanced resistance has allowed the Co-based implants to be employed for artificial hip joints wherein wear due to direct contact between the bone plate and the femoral head may result. The combination of high strength and durability is one of the major advantages for the Co-Cr-Mo implants and therefore they are used widely [54]. The elastic modulus of Co-Cr alloy based implants is also relatively higher than the pure Ti or Ti alloys [55, 56].

The Co-Cr alloy-based implants have higher density and elastic modulus in comparison to bone [57]. These characteristics allow for greater stress shielding than that of Ti and Ti alloy-based implants [58]. However, the Co-Cr alloy-based implants have lower biocompatibility and osseointegration capacity than the Ti-based implants. Therefore, Ti-based implants are mainly used for the elements that make direct contact with the bone, whereas the elements such as the rods in spinal fixation that do not interfere with the bone directly are made using Co-Cr-based alloys. It has been, however, revealed that metal corrosion and shredding may result at the site making contact with the Ti or Co implants, due to increasing frictional load. The tissues near the metallic implants experience metallosis, as in the case of total hip arthroplasty, spinal fixation, and knee implants.

The excessive stiffness of the structure in the case of Co-Cr alloy-based implants for the 3D technique to be implemented can be overcome by the incorporation of nano- and microstructures within the bulk geometry. The incorporation and simultaneous control over the nano- and microstructures can result in implants with reduced elastic modulus. Further, the discrepancy of stiffness between the alloyed implants and the bone can also be eliminated using bulk modifications. Co-Cr-based implants with the desired pore architecture and macrogeometry can be produced using the electron beam machining (EBM) technique [59]. The implants showed around 28% contact with the bone when implanted into adult sheep femora for around 26 weeks. Although this contact percentage was lower than that observed for Ti-6Al-4V implants, the apatite-collagen ratio, densification, and phosphate-carbonate ratio was similar for the Ti and Co-Cr alloy-based implants. Co-Cr porous structures also showed higher osteocyte density at their periphery [59].

Meshed and solid implants can be printed using EBM in addition to foam monoliths. Architecture of $Cr_{23}C_6$ precipitates in the columnar directional form with spacing of around 2 μm has been observed for the foam ligaments and mesh struts from Co-29Cr-6Mo alloys. The microstructures reveal the directional solidification of the solid cylindrical components in the assembly process [60].

Fine cellular microstructures in CoCrMo implants have been observed when employing selective laser melting (SLM) for the fabrication of Co-based implants. The fine cellular microstructure was a result of subsequent rapid cooling and strong temperature gradients maintained during the melting of the alloy. The grain boundaries were observed to be depleted in Co and enriched in Mo. Such processing steps result in reduced formation of a martensitic ε phase and minimize the precipitation of carbides at the surface. Further, a highly corrosion-resistant 3D printed implant results from using the SLM fabrication technique [61].

4.2.4 Tantalum-based bioimplants

Tantalum has been investigated for many biomaterial applications owing to its excellent biocompatibility characteristics. The biocompatibility property has aimed tantalum-based bioimplants for environments where biocompatibility and corrosion resistance are the major requirements [62]. A Ta_2O_5 protective film on the implant surface provides anticorrosion characteristics to tantalum-based bioimplants [63, 64]. Tantalum is one of the attractive materials for artificial joints because of its excellent bone-bonding characteristics. These can either be used for bulk materials or as a coating on titanium and stainless steel bioimplants, lending them better osseointegration and corrosion resistance [65–68].

The higher elastic modulus as well as the difficulty in fabrication with higher precision are the major limitations to the wider use of tantalum. The elastic modulus of tantalum is above 186 GPa and the density is around 16.6 g/cm^3. The elastic modulus is very high in comparison to cancellous bone (0.1–0.45 GPa) and cortical (10–16 GPa) and, therefore, because of this huge mismatch, the tantalum metal is detrimental [69, 70]. Further, a high melting temperature of around 3020°C is a major challenge for the bulk production of bioimplants based on tantalum. Patients have been observed to suffer from headaches because of the high conductivity of tantalum when used for cranioplasty [71].

Implants have been manufactured with 50 wt% of Ta and Ti using the SLM manufacturing technique. The Ta nanoparticles were randomly dispersed in the Ti-Ta matrix. Equiaxed grains of β phase Ti and Ta composed the matrix with a random orientation. This results in a β stabilizing effect and rapid solidification [72]. The fabricated Ti-Ta was revealed to possess a higher strength-to-modulus ratio than pure Ti and Ti-6Al-4V alloys.

4.3 Biodegradable metals

Biodegradable metals are one of the attractive alternatives to implants made from permanent metals for fracture fixations in cases where complete tissue regeneration is desired. Zinc (Zn), iron (Fe) and magnesium (Mg) are some of the best-explored biodegradable metals, used extensively for cardiovascular and orthopedic applications [73]. These biodegradable metals offer excellent in vivo biocompatibility, have the required mechanical strength to provide support to bone undergoing a regeneration process, and have a controlled degradation profile. The mechanical behavior of bioresorbable metals has been found to be far superior to the bioresorbable polymers [74]. The products of biodegradation of the biodegradable biometals are metabolized by the host cells, unlike the byproducts of polymeric products, which result in necrosis and inflammatory issues [75].

4.3.1 *Magnesium-based biodegradable implants*

Magnesium possesses high specific strength among the other candidate biodegradable materials [76] and also has relatively lower density and elastic modulus. The elastic modulus is around 41 GPa and the density is 1.75 g/cm^3. These properties are very close to those of bone and therefore aid in minimization of risk pertaining to stress shielding [77]. The Mg-based implants such as rods, plates, and screws provide for the required mechanical strength and degrade gradually, thereby providing space for the regeneration of the bone tissue (https://www.accessscience.com:443/content/biodegradable-metal-implants/BR0601151). The gradual degradation leads to positive contribution to the strength and density of the bone, as there is a gradual increase in the force from the implant onto the bone. The biodegradable implant eventually will be fully degraded and thereby minimizes the metal-related sensitivity issues that occur with Co- or Ti-based implants.

The Mg-based biodegradable implants, however, have a rapid rate of corrosion, especially in environments full of biological fluids [78]. The rapid corrosion rate results in release of large Mg^{2+} ions, leading to untimely loss in the mechanical strength of the implants [79–81]. The release of hydrogen gas is also very rapid along with the corrosion rate, which may lead to formation of gas pockets around the implant [82]. Thus an extremely higher corrosion rate may hinder the regeneration of the bone. The degradation kinetics of Mg can be controlled via alloying.

The solid free-form (SFF) fabrication technique has been used in preparation of Mg scaffolds with ordered pore macrostructures. The use of SFF has resulted in higher resolution in terms of volume fraction, stiffness, and porosity [83]. Pore dimensions of 6% and 9% and surface roughness of around 11 μm have been achieved with the SFF technology [83, 84]. The surface area achieved was high enough for efficient interactions between cells and surface.

4.3.2 *Zinc-based biodegradable implants*

Zinc is one of the important elements that play a crucial role in the functioning of proteins as well as the related structures. It is an essential element for the proper catalytic functioning of more than 400 enzymes, including stabilization and folding of protein subdomains. Examples include proteins engaged in replication of nucleic acid, RNA polymerases, and DNA-binding of the factors related to eukaryotic transcription [85]. Since the environment of stent operation is such that most tissues in proximity have good tolerances to Zn ions, Zn is being explored for applications pertaining to bioresorbable metallic stents. The main processes responsible for the corrosion of zinc are the cathodic reduction of the oxygen dissolved and the anodic dissolution [75]. The corrosion rate is also affected with the pH level of the surrounding environment. The corrosion rate of Zn is comparatively lower than that of pure Mg.

Zn-based biodegradable stents do not lead to any inflammatory responses, which was evident when Zn stents were used for rat arteries. Zn and its alloys have also been explored for fixation of fractures, besides their applications for stents. Enhanced corrosion resistance as well as improved mechanical properties are obtained by combining Zn with Mg [73].

4.3.3 Iron-based biodegradable implants

Fe also undergoes local corrosion in the presence of dissolved oxygen. The tendency of iron to dissolve is the lowest in comparison to the other biodegradable metals. The formation of a protective oxide layer over the surface is one of the major reasons for lower corrosion rates. The oxide layer on the surface of Fe acts as a barrier to the rapid degradation phenomenon. The investigations into usage of Fe-based stents have shown no inflammatory responses or thrombosis or excessive toxicity. High radial strength is one of the characteristic properties of Fe stents, which aids in fabrication of extremely thin stents. The ductility of the Fe stents is also higher and hence deployment into the arteries is much easier [75].

However, with the advantages are associated certain limitations that result in undesirable consequences. For instance, incomplete corrosion has been revealed for Fe stents. To avoid this, alloying has been suggested as an alternative, to increase the rate of degradation. However, control of the formation of by-products of iron oxide degradation as well as the influx of metallic ions into the surrounding tissues is required. The increased quantity of Fe in the system can lead to inflammation, damage to lipid membranes, an increment in free radicals, and damage to DNA and proteins [1].

Further, Fe-based bioimplants may interfere with magnetic resonance imaging (MRI) because of the magnetic nature of iron. A strong magnetic field may cause tremendous heating of the stents, resulting in change in shape as well as position of the stents. However, such effects have positive impacts that if harnessed could enhance the therapeutic outcome of any treatment. For instance, osseous tissue generation can be stimulated by employing porous scaffolds with iron nanoparticles. The intracellular pathways associated with bone formation are activated as a result of application of an external magnetic field [86]. Triggering of drug release can also be accomplished using external magnetic fields. The heat generated locally can be used for thermal therapy of cancer and also could treat any implant-associated infections.

4.4 Limitations of biomaterials

Metallic biomaterials are being used widely for the fabrication of different prosthetic parts. Thus it is essential to have knowledge of the limitations associated with these biomaterials. Also it is essential to understand the different strategies that could be adopted to overcome or minimize these limitations. The present section illuminates some of the limitations and the different strategies for overcoming such limitations.

4.4.1 Biocompatibility

Both permanent and biodegradable metal-based implants are biocompatible; the biocompatibility is negatively affected by untimely degradation and excessive wear. This may lead to hindrances in healing and hence to long-term damage [87]. The biocompatibility may also be reduced as a result of techniques used for enhancing the mechanical properties as well as the corrosion resistance of the material. For instance, inflammation in the tissues surrounding the implant increases with the addition of Ni or Ti [88]. Such issues can, however, be minimized to some extent by avoiding direct contact between the implants and the cells surrounding the implants. For instance, Co-Cr based rods can be affixed to the bone using biocompatible Ti screws [89]. Interactions between the host cells and the surface are also affected when employing strategies for preventing bacterial contamination. Using such strategies can also hinder osseointegration [90, 91].

The process of osseointegration as well as that of cell attachment is stimulated by additional surface roughness and porosity. The strategies for achieving such modifications range from sandblasting to laser ablation. For example, the required surface roughness over the implants can be produced by sandblasting Ti with large grit acid. This provides a suitable environment for growth and attachment of cells [92]. The surface modification strategies can directly affect the manner in which the surfaces and pathogenic microorganisms interact with each other [93, 94].

Biocompatibility can also be increased by bulk modifications such as those of alloying elements that stimulate for cell attachment and proliferation, introduction of porosity in a controlled manner, or changing the structure of the grains. Complex modifications such as providing complex interconnected architectures and channels have been made possible with the promising and emerging technology of 3D printing [95, 96]. The degradation profile of the material can be altered by increasing the surface area in contact with the corrosive implant surroundings. However, this may lead to development of sarcoma as result of substantial release of ions from the implants [97]. This can be minimized by using a thin film of polymer or ceramic as a coating on the surface of the 3D scaffolds.

A suitable modification technique must be selected that doesn't compromise the required dimensions and at the same time ensures connectivity of internal channels and pores. For metal scaffolds, techniques such as solvent casting or spray-, dip-, or spin-coatings can be used. These techniques produce a thin layer of a film that may suffer from nonuniformity of surface roughness, chemical functionalities, etc. [98]. The fine internal structures of the implants can be affected negatively with a nonuniform film thickness.

4.4.2 Surface colonization and formation of biofilm

The implants that promote host cell-surface modifications and simultaneously prevent the formation of biofilm are highly desired. The attachment of protein takes place on introducing an implant into the body of a patient, along with the colonization of the surface by the host and pathogenic cells [99]. In certain instances in which this colonization doesn't occur, long-term antimicrobial strategies are followed to prevent the formation of a biofilm [100, 101]. One of the strategies for the antibacterial coating includes the use of silver nanoparticles for Ti-based bioimplants.

One of the emerging and promising technological fields is plasma-based biomedical technologies, which have been used widely for decontamination at low temperatures in the food industry and for tissue regeneration [102–106]. Cellular activities are significantly altered when the cells are treated with plasma. This results in controlling the processes responsible for the formation of biofilm, carcinogenesis, and tissue regeneration [102, 103]. Plasma-enabled biomedical technologies are also used for carrying out long-lasting controlled surface modifications [93, 94]. This also includes chemical functionalization and deposition of films with antibacterial properties [107–109]. Plasma has also been used for the deposition of surface nanostructures of highly complex nature from a wide range of materials [17, 18, 110] that can aid in a wider control over the attachment behavior of microorganisms and cells.

Chemical gradients have been successfully obtained in 3D porous scaffolds using the technique of plasma deposition, resulting in enhanced cell viability in comparison to untreated materials [111–113]. It is also possible to add the different N- or O-functional groups and their combinations and to deposit nano- and microscale features throughout the porous structure [114].

4.5 Conclusions

The present work reviews the currently used biomaterials with their limitations and the strategies that have been adopted for their modifications. However, the need for new or modified materials can only address the challenges that are associated with the current implants. The biomaterials should not only minimize the medical complications but also should be aesthetically comfortable for the patients. Significant values can be associated with future biomaterials using the current material engineering techniques. The next-generation implants are to be manufactured using nanomaterials that can serve multiple purposes.

References

[1] D. Hong, D.T. Chou, O.I. Velikokhatnyi, A. Roy, B. Lee, I. Swink…P.N. Kumta, Binder-jetting 3D printing and alloy development of new biodegradable Fe-Mn-Ca/Mg alloys, Acta Biomater. 45 (2016) 375–386.

[2] O.O. Ige, L.E. Umoru, S. Aribo, Natural products: a minefield of biomaterials, ISRN Mater. Sci. 2012 (2012).

[3] J.Z. Lu, L.J. Wu, G.F. Sun, K.Y. Luo, Y.K. Zhang, J. Cai...X.M. Luo, Microstructural response and grain refinement mechanism of commercially pure titanium subjected to multiple laser shock peening impacts, Acta Mater. 127 (2017) 252–266.
[4] P. Trivedi, K.C. Nune, R.D.K. Misra, S. Goel, R. Jayganthan, A. Srinivasan, Grain refinement to submicron regime in multiaxial forged Mg-2Zn-2Gd alloy and relationship to mechanical properties, Mater. Sci. Eng. A 668 (2016) 59–65.
[5] S. Bhat, A. Kumar, Biomaterials and bioengineering tomorrow's healthcare, Biomatter 3 (3) (2013) e24717.
[6] J. Henkel, M.A. Woodruff, D.R. Epari, R. Steck, V. Glatt, I.C. Dickinson... D.W. Hutmacher, Bone regeneration based on tissue engineering conceptions—a 21st century perspective, Bone Res. 1 (2013) 216–248.
[7] N. Hort, Y. Huang, D. Fechner, M. Störmer, C. Blawert, F. Witte...F. Feyerabend, Magnesium alloys as implant materials—principles of property design for Mg–RE alloys, Acta Biomater. 6 (5) (2010) 1714–1725.
[8] M.P. Staiger, A.M. Pietak, J. Huadmai, G. Dias, Magnesium and its alloys as orthopedic biomaterials: a review, Biomaterials 27 (9) (2006) 1728–1734.
[9] H. Wang, Y. Estrin, H. Fu, G. Song, Z. Zúberová, The effect of pre-processing and grain structure on the bio-corrosion and fatigue resistance of magnesium alloy AZ31, Adv. Eng. Mater. 9 (11) (2007) 967–972.
[10] J. Zhang, C. Xu, Y. Jing, S. Lv, S. Liu, D. Fang...R. Wu, New horizon for high performance Mg-based biomaterial with uniform degradation behavior: formation of stacking faults, Sci. Rep. 5 (2015) 13933.
[11] E.P. Ivanova, K. Bazaka, R.J. Crawford, New Functional Biomaterials for Medicine and Healthcare, Woodhead Publishing, Cambridge, UK, 2014.
[12] A. Srivastav, An overview of metallic biomaterials for bone support and replacement, in: Biomedical Engineering, Trends in Materials Science, InTech, 2011.
[13] L. Pacifici, F. De Angelis, A. Orefici, A. Cielo, Metals used in maxillofacial surgery, ORAL Implantol. 9 (Suppl. 1/2016 to N 4/2016) (2016) 107.
[14] M. Saini, Y. Singh, P. Arora, V. Arora, K. Jain, Implant biomaterials: a comprehensive review, World J. Clin. Cases 3 (1) (2015) 52.
[15] M. Niinomi, M. Nakai, Titanium-based biomaterials for preventing stress shielding between implant devices and bone, Int. J. Biomater. 2011 (2011).
[16] K. Bazaka, M.V. Jacob, Implantable devices: issues and challenges, Electronics 2 (1) (2012) 1–34.
[17] K. Bazaka, M.V. Jacob, K. Ostrikov, Sustainable life cycles of natural-precursor-derived nanocarbons, Chem. Rev. 116 (1) (2015) 163–214.
[18] K. Bazaka, M.V. Jacob, W. Chrzanowski, K. Ostrikov, Anti-bacterial surfaces: natural agents, mechanisms of action, and plasma surface modification, RSC Adv. 5 (60) (2015) 48739–48759.
[19] M. Cappiello, R. Luongo, D. Di Iorio, C. Bugea, R. Cocchetto, R. Celletti, Evaluation of peri-implant bone loss around platform-switched implants, Int. J. Periodontics Restorative Dent. 28 (4) (2008) 347–355.
[20] K. Prasad, O. Bazaka, M. Chua, M. Rochford, L. Fedrick, J. Spoor...D. Markwell, Metallic biomaterials: current challenges and opportunities, Materials 10 (8) (2017) 884.
[21] L.D. Campelo, J.R.D. Camara, Flapless implant surgery: a 10-year clinical retrospective analysis, Int. J. Oral Maxillofac. Implants 17 (2) (2002).
[22] K. Bazaka, M.V. Jacob, R.J. Crawford, E.P. Ivanova, Efficient surface modification of biomaterial to prevent biofilm formation and the attachment of microorganisms, Appl. Microbiol. Biotechnol. 95 (2) (2012) 299–311.

[23] G.A. Van der Weijden, K.M. Van Bemmel, S. Renvert, Implant therapy in partially edentulous, periodontally compromised patients: a review, J. Clin. Periodontol. 32 (5) (2005) 506–511.
[24] S.V. Murphy, A. Atala, 3D bioprinting of tissues and organs, Nat. Biotechnol. 32 (8) (2014) 773.
[25] K. Jakab, C. Norotte, F. Marga, K. Murphy, G. Vunjak-Novakovic, G. Forgacs, Tissue engineering by self-assembly and bio-printing of living cells, Biofabrication 2 (2) (2010) 022001.
[26] V. Mironov, N. Reis, B. Derby, Bioprinting: a beginning, Tissue Eng. 12 (4) (2006) 631–634.
[27] S.C. Mears, S.L. Kates, A guide to improving the care of patients with fragility fractures, edition 2, Geriatr. Orthop. Surg. Rehabil. 6 (2) (2015) 58–120.
[28] H.K. Uhthoff, P. Poitras, D.S. Backman, Internal plate fixation of fractures: short history and recent developments, J. Orthop. Sci. 11 (2) (2006) 118–126.
[29] W.H. Hatfield, Rustless steels as applied to automobiles and aircraft, Proc. Inst. Automobile Eng. 25 (2) (1931) 285–304.
[30] P. Ducheyne, D.H. Kohn, Materials Science and Technology—A Comprehensive Treatment, Medical and Dental Materials, Weinheim Vol. 14, 1998, pp. 39–41.
[31] P. Driscoll, Materials Used in Stent Construction, Biddeford, ME, MedMarket Diligence, LLC, 2009.
[32] M. Sumita, Present status and future trend of metallic materials used in orthopedics, Orthop. Surg. 48 (1997) 927–934.
[33] M. Frame, Low cost orthopaedic implant trials created using 3D printing technology, Bone Joint J. 98 (Supp. 1) (2016) 134.
[34] J.P. Kruth, B. Vandenbroucke, J. Van Vaerenbergh, I. Naert, in: Rapid manufacturing of dental prostheses by means of selective laser sintering/melting, Proceedings of the AFPR S, 4, 2005, pp. 176–186.
[35] D.C. Hansen, Metal corrosion in the human body: the ultimate bio-corrosion scenario, Electrochem. Soc. Interface 17 (2) (2008) 31.
[36] O. Mouzin, K. Søballe, J.E. Bechtold, Loading improves anchorage of hydroxyapatite implants more than titanium implants, J. Biomed. Mater. Res. A 58 (1) (2001) 61–68.
[37] X. Zhao, M. Niinomi, M. Nakai, T. Ishimoto, T. Nakano, Development of high Zr-containing Ti-based alloys with low Young's modulus for use in removable implants, Mater. Sci. Eng. C 31 (7) (2011) 1436–1444.
[38] H. Nakada, T. Sakae, Y. Tanimoto, M. Teranishi, T. Kato, T. Watanabe…R.Z. LeGeros, Assessment of the quality of newly formed bone around titanium alloy implants by using X-ray photoelectron spectroscopy, Int. J. Biomater. 2012 (2012).
[39] S.L.M. Ribeiro Filho, C.H. Lauro, A.H.S. Bueno, L.C. Brandão, Effects of the dynamic tapping process on the biocompatibility of Ti-6Al-4V alloy in simulated human body environment, Arab. J. Sci. Eng. 41 (11) (2016) 4313–4326.
[40] I. Levchenko, K.K. Ostrikov, J. Zheng, X. Li, M. Keidar, K.B. Teo, Scalable graphene production: perspectives and challenges of plasma applications, Nanoscale 8 (20) (2016) 10511–10527.
[41] B. Lethaus, P. Kessler, R. Boeckman, L.J. Poort, R. Tolba, Reconstruction of a maxillary defect with a fibula graft and titanium mesh using CAD/CAM techniques, Head Face Med 6 (1) (2010) 16.
[42] K. Malhotra, A. Sharma, G. Giraddi, A.K. Shahi, Versatility of titanium 3D plate in comparison with conventional titanium miniplate fixation for the management of mandibular fracture, J. Maxillofac. Oral Surg. 11 (3) (2012) 284–290.

[43] J. Li, P. Li, H. Lu, L. Shen, W. Tian, J. Long, W. Tang, Digital design and individually fabricated titanium implants for the reconstruction of traumatic zygomatico-orbital defects, J. Craniofac. Surg. 24 (2) (2013) 363–368.
[44] H. Fan, J. Fu, X. Li, Y. Pei, X. Li, G. Pei, Z. Guo, Implantation of customized 3-D printed titanium prosthesis in limb salvage surgery: a case series and review of the literature, World J. Surg. Oncol. 13 (1) (2015) 308.
[45] S. Ponader, C. Von Wilmowsky, M. Widenmayer, R. Lutz, P. Heinl, C. Körner… K.A. Schlegel, In vivo performance of selective electron beam-melted Ti-6Al-4V structures, J. Biomed. Mater. Res. A 92 (1) (2010) 56–62.
[46] F. Mangano, F. Luongo, J.A. Shibli, S. Anil, C. Mangano, Maxillary overdentures supported by four splinted direct metal laser sintering implants: a 3-year prospective clinical study, Int. J. Dent. 2014 (2014).
[47] F. Mangano, S. Pozzi-Taubert, P.A. Zecca, G. Luongo, R.L. Sammons, C. Mangano, Immediate restoration of fixed partial prostheses supported by one-piece narrow-diameter selective laser sintering implants: a 2-year prospective study in the posterior jaws of 16 patients, Implant. Dent. 22 (4) (2013) 388–393.
[48] T. Traini, C. Mangano, R.L. Sammons, F. Mangano, A. Macchi, A. Piattelli, Direct laser metal sintering as a new approach to fabrication of an isoelastic functionally graded material for manufacture of porous titanium dental implants, Dent. Mater. 24 (11) (2008) 1525–1533.
[49] G.E. Ryan, A.S. Pandit, D.P. Apatsidis, Porous titanium scaffolds fabricated using a rapid prototyping and powder metallurgy technique, Biomaterials 29 (27) (2008) 3625–3635.
[50] C. Mangano, M. Raspanti, T. Traini, A. Piattelli, R. Sammons, Stereo imaging and cytocompatibility of a model dental implant surface formed by direct laser fabrication, J. Biomed. Mater. Res. A 88 (3) (2009) 823–831.
[51] S. Tunchel, A. Blay, R. Kolerman, E. Mijiritsky, J.A. Shibli, 3D printing/additive manufacturing single titanium dental implants: a prospective multicenter study with 3 years of follow-up, Int. J. Dent. 2016 (2016).
[52] C. Mangano, A. De Rosa, V. Desiderio, R. d'Aquino, A. Piattelli, F. De Francesco… G. Papaccio, The osteoblastic differentiation of dental pulp stem cells and bone formation on different titanium surface textures, Biomaterials 31 (13) (2010) 3543–3551.
[53] L. Witek, C. Marin, R. Granato, E.A. Bonfante, F. Campos, J. Bisinotto…P.G. Coelho, Characterization and in vivo evaluation of laser sintered dental endosseous implants in dogs, J. Biomed. Mater. Res. B Appl. Biomater. 100 (6) (2012) 1566–1573.
[54] M. Niinomi, Recent metallic materials for biomedical applications, Metall. Mater. Trans. A 33 (3) (2002) 477.
[55] A. Aherwar, A.K. Singh, A. Patnaik, Cobalt based alloy: a better choice biomaterial for hip implants, Trends Biomater. Artif. Organs 30 (1) (2016) 50–55.
[56] S.J. Li, R. Yang, S. Li, Y.L. Hao, Y.Y. Cui, M. Niinomi, Z.X. Guo, Wear characteristics of Ti–Nb–Ta–Zr and Ti–6Al–4V alloys for biomedical applications, Wear 257 (9–10) (2004) 869–876.
[57] Y. Li, C. Yang, H. Zhao, S. Qu, X. Li, Y. Li, New developments of Ti-based alloys for biomedical applications, Materials 7 (3) (2014) 1709–1800.
[58] S. Nayak, B. Bhushan, R. Jayaganthan, P. Gopinath, R.D. Agarwal, D. Lahiri, Strengthening of Mg based alloy through grain refinement for orthopaedic application, J. Mech. Behav. Biomed. Mater. 59 (2016) 57–70.
[59] F.A. Shah, O. Omar, F. Suska, A. Snis, A. Matic, L. Emanuelsson…A. Palmquist, Long-term osseointegration of 3D printed CoCr constructs with an interconnected open-pore architecture prepared by electron beam melting, Acta Biomater. 36 (2016) 296–309.

[60] L.E. Murr, K.N. Amato, S.J. Li, Y.X. Tian, X.Y. Cheng, S.M. Gaytan…R.B. Wicker, Microstructure and mechanical properties of open-cellular biomaterials prototypes for total knee replacement implants fabricated by electron beam melting, J. Mech. Behav. Biomed. Mater. 4 (7) (2011) 1396–1411.
[61] Y.S. Hedberg, B. Qian, Z. Shen, S. Virtanen, I.O. Wallinder, In vitro biocompatibility of CoCrMo dental alloys fabricated by selective laser melting, Dent. Mater. 30 (5) (2014) 525–534.
[62] R.S. Namur, K.M. Reyes, C.E.B. Marino, Growth and electrochemical stability of compact tantalum oxides obtained in different electrolytes for biomedical applications, Mater. Res. 18 (2015) 91–97.
[63] D. Cristea, I. Ghiuta, D. Munteanu, Tantalum based materials for implants and prostheses applications, Bull. Transilvania Univ. Brasov Ser. I Eng. Sci. 8 (2) (2015) 151.
[64] N. Wang, H. Li, J. Wang, S. Chen, Y. Ma, Z. Zhang, Study on the anticorrosion, biocompatibility, and osteoinductivity of tantalum decorated with tantalum oxide nanotube array films, ACS Appl. Mater. Interfaces 4 (9) (2012) 4516–4523.
[65] V.K. Balla, S. Banerjee, S. Bose, A. Bandyopadhyay, Direct laser processing of a tantalum coating on titanium for bone replacement structures, Acta Biomater. 6 (6) (2010) 2329–2334.
[66] M. Roy, V.K. Balla, A. Bandyopadhyay, S. Bose, MgO-doped tantalum coating on Ti: microstructural study and biocompatibility evaluation, ACS Appl. Mater. Interfaces 4 (2) (2012) 577–580.
[67] P. Sevilla, C. Aparicio, J.A. Planell, F.J. Gil, Comparison of the mechanical properties between tantalum and nickel–titanium foams implant materials for bone ingrowth applications, J. Alloys Compd. 439 (1–2) (2007) 67–73.
[68] M. Stiehler, M. Lind, T. Mygind, A. Baatrup, A. Dolatshahi-Pirouz, H. Li…C. Bünger, Morphology, proliferation, and osteogenic differentiation of mesenchymal stem cells cultured on titanium, tantalum, and chromium surfaces, J. Biomed. Mater. Res. A 86 (2) (2008) 448–458.
[69] Y. Liu, C. Bao, D. Wismeijer, G. Wu, The physicochemical/biological properties of porous tantalum and the potential surface modification techniques to improve its clinical application in dental implantology, Mater. Sci. Eng. C 49 (2015) 323–329.
[70] T. Lu, J. Wen, S. Qian, H. Cao, C. Ning, X. Pan…P.K. Chu, Enhanced osteointegration on tantalum-implanted polyetheretherketone surface with bone-like elastic modulus, Biomaterials 51 (2015) 173–183.
[71] A.M. Shah, H. Jung, S. Skirboll, Materials used in cranioplasty: a history and analysis, Neurosurg. Focus. 36 (4) (2014) E19.
[72] S.L. Sing, W.Y. Yeong, F.E. Wiria, Selective laser melting of titanium alloy with 50 wt% tantalum: microstructure and mechanical properties, J. Alloys Compd. 660 (2016) 461–470.
[73] M. Heiden, E. Walker, L. Stanciu, Magnesium, iron and zinc alloys, the trifecta of bioresorbable orthopaedic and vascular implantation—a review, J. Biotechnol. Biomater. 5 (2) (2015) 1.
[74] D. Zhao, F. Witte, F. Lu, J. Wang, J. Li, L. Qin, Current status on clinical applications of magnesium-based orthopaedic implants: a review from clinical translational perspective, Biomaterials 112 (2017) 287–302.
[75] P.K. Bowen, E.R. Shearier, S. Zhao, R.J. Guillory, F. Zhao, J. Goldman, J.W. Drelich, Biodegradable metals for cardiovascular stents: from clinical concerns to recent Zn-alloys, Adv. Healthc. Mater. 5 (10) (2016) 1121–1140.

[76] D. Persaud-Sharma, A. McGoron, Biodegradable magnesium alloys: a review of material development and applications, J. Biomim. Biomater. Tissue Eng. 12 (2011) 25–39. Trans Tech Publications.
[77] T.S. Shih, W.S. Liu, Y.J. Chen, Fatigue of as-extruded AZ61A magnesium alloy, Mater. Sci. Eng. A 325 (1–2) (2002) 152–162.
[78] H.S. Brar, M.O. Platt, M. Sarntinoranont, P.I. Martin, M.V. Manuel, Magnesium as a biodegradable and bioabsorbable material for medical implants, JOM 61 (9) (2009) 31–34.
[79] A. Atrens, M. Liu, N.I.Z. Abidin, Corrosion mechanism applicable to biodegradable magnesium implants, Mater. Sci. Eng. B 176 (20) (2011) 1609–1636.
[80] K. Bazaka, N. Ketheesan, M.V. Jacob, Polymer encapsulation of magnesium to control biodegradability and biocompatibility, J. Nanosci. Nanotechnol. 14 (10) (2014) 8087–8093.
[81] N.T. Kirkland, N. Birbilis, M.P. Staiger, Assessing the corrosion of biodegradable magnesium implants: a critical review of current methodologies and their limitations, Acta Biomater. 8 (3) (2012) 925–936.
[82] G. Song, S. Song, A possible biodegradable magnesium implant material, Adv. Eng. Mater. 9 (4) (2007) 298–302.
[83] T.L. Nguyen, M.P. Staiger, G.J. Dias, T.B. Woodfield, A novel manufacturing route for fabrication of topologically-ordered porous magnesium scaffolds, Adv. Eng. Mater. 13 (9) (2011) 872–881.
[84] M.P. Staiger, I. Kolbeinsson, N.T. Kirkland, T. Nguyen, G. Dias, T.B. Woodfield, Synthesis of topologically-ordered open-cell porous magnesium, Mater. Lett. 64 (23) (2010) 2572–2574.
[85] J.E. Coleman, Zinc proteins: enzymes, storage proteins, transcription factors, and replication proteins, Annu. Rev. Biochem. 61 (1) (1992) 897–946.
[86] A. Ortolani, M. Bianchi, M. Mosca, S. Caravelli, M. Fuiano, M. Marcacci, A. Russo, The prospective opportunities offered by magnetic scaffolds for bone tissue engineering: a review, Joints 4 (4) (2016) 228.
[87] J. Autian, Biological model systems for the testing of the toxicity of biomaterials, in: Polymers in Medicine and Surgery, Springer, Boston, MA, 1975, pp. 181–203.
[88] T.P. Chaturvedi, Corrosive behaviour of implant biomaterials in oral environment, Mater. Technol. 31 (12) (2016) 689–695.
[89] J.J. Jacobs, A.K. Skipor, P.F. Doorn, P. Campbell, T.P. Schmalzried, J. Black, H.C. Amstutz, Cobalt and chromium concentrations in patients with metal on metal total hip replacements, Clin. Orthop. Relat. Res. 329 (1996) S256–S263.
[90] D. Campoccia, L. Montanaro, C.R. Arciola, The significance of infection related to orthopedic devices and issues of antibiotic resistance, Biomaterials 27 (11) (2006) 2331–2339.
[91] R.O. Darouiche, Treatment of infections associated with surgical implants, N. Engl. J. Med. 350 (14) (2004) 1422–1429.
[92] R.Z. Valiev, I.P. Semenova, V.V. Latysh, H. Rack, T.C. Lowe, J. Petruzelka…J. Sochová, Nanostructured titanium for biomedical applications, Adv. Eng. Mater. 10 (8) (2008).
[93] K. Bazaka, R.J. Crawford, E.P. Ivanova, Do bacteria differentiate between degrees of nanoscale surface roughness? Biotechnol. J. 6 (9) (2011) 1103–1114.
[94] K. Bazaka, M.V. Jacob, R.J. Crawford, E.P. Ivanova, Plasma-assisted surface modification of organic biopolymers to prevent bacterial attachment, Acta Biomater. 7 (5) (2011) 2015–2028.

[95] H.N. Chia, B.M. Wu, Recent advances in 3D printing of biomaterials, J. Biol. Eng. 9 (1) (2015) 4.
[96] S.J. Hollister, Porous scaffold design for tissue engineering, Nat. Mater. 4 (7) (2005) 518.
[97] S.F.S. Shirazi, S. Gharehkhani, M. Mehrali, H. Yarmand, H.S.C. Metselaar, N.A. Kadri, N.A.A. Osman, A review on powder-based additive manufacturing for tissue engineering: selective laser sintering and inkjet 3D printing, Sci. Technol. Adv. Mater. 16 (3) (2015) 033502.
[98] J. Vaithilingam, S. Kilsby, R.D. Goodridge, S.D. Christie, S. Edmondson, R.J. Hague, Functionalisation of Ti6Al4V components fabricated using selective laser melting with a bioactive compound, Mater. Sci. Eng. C 46 (2015) 52–61.
[99] V.T. Pham, V.K. Truong, A. Orlowska, S. Ghanaati, M. Barbeck, P. Booms… C.J. Kirkpatrick, "Race for the surface": eukaryotic cells can win, ACS Appl. Mater. Interfaces 8 (34) (2016) 22025–22031.
[100] V. Antoci Jr., C.S. Adams, J. Parvizi, H.M. Davidson, R.J. Composto, T.A. Freeman… N.J. Hickok, The inhibition of *Staphylococcus epidermidis* biofilm formation by vancomycin-modified titanium alloy and implications for the treatment of periprosthetic infection, Biomaterials 29 (35) (2008) 4684–4690.
[101] S. Fujimura, T. Sato, S. Hayakawa, M. Kawamura, E. Furukawa, A. Watanabe, Antimicrobial efficacy of combined clarithromycin plus daptomycin against biofilms-formed methicillin-resistant *Staphylococcus aureus* on titanium medical devices, J. Infect. Chemother. 21 (10) (2015) 756–759.
[102] M. Ishaq, K. Bazaka, K. Ostrikov, Pro-apoptotic NOXA is implicated in atmospheric-pressure plasma-induced melanoma cell death, J. Phys. D. Appl. Phys. 48 (46) (2015) 464002.
[103] M. Ishaq, K. Bazaka, K. Ostrikov, Intracellular effects of atmospheric-pressure plasmas on melanoma cancer cells, Phys. Plasma. 22 (12) (2015) 122003.
[104] M. Ishaq, A. Rowe, K. Bazaka, M. Krockenberger, M.D. Evans, K.K. Ostrikov, Effect of atmospheric-pressure plasmas on drug resistant melanoma: the challenges of translating in vitro outcomes into animal models, Plasma Med. 6 (1) (2016) 67–83.
[105] X.Q. Wang, F.P. Wang, W. Chen, J. Huang, K. Bazaka, K.K. Ostrikov, Non-equilibrium plasma prevention of *Schistosoma japonicum* transmission, Sci. Rep. 6 (2016) 35353.
[106] R. Zhou, R. Zhou, X. Zhang, J. Li, X. Wang, Q. Chen, K.K. Ostrikov, Synergistic effect of atmospheric-pressure plasma and TiO 2 photocatalysis on inactivation of *Escherichia coli* cells in aqueous media, Sci. Rep. 6 (2016) 39552.
[107] J. Ahmad, K. Bazaka, M.V. Jacob, Optical and surface characterization of radio frequency plasma polymerized 1-isopropyl-4-methyl-1, 4-cyclohexadiene thin films, Electronics 3 (2) (2014) 266–281.
[108] M.V. Jacob, K. Bazaka, D. Taguchi, T. Manaka, M. Iwamoto, Electron-blocking hole-transport polyterpenol thin films, Chem. Phys. Lett. 528 (2012) 26–28.
[109] M.V. Jacob, N.S. Olsen, L.J. Anderson, K. Bazaka, R.A. Shanks, Plasma polymerised thin films for flexible electronic applications, Thin Solid Films 546 (2013) 167–170.
[110] M.V. Jacob, R.S. Rawat, B. Ouyang, K. Bazaka, D.S. Kumar, D. Taguchi… O.K. Varghese, Catalyst-free plasma enhanced growth of graphene from sustainable sources, Nano Lett. 15 (9) (2015) 5702–5708.
[111] F. Intranuovo, R. Gristina, F. Brun, S. Mohammadi, G. Ceccone, E. Sardella… P. Favia, Plasma modification of PCL porous scaffolds fabricated by solvent-casting/particulate-leaching for tissue engineering, Plasma Process. Polym. 11 (2) (2014) 184–195.

[112] L. Safinia, N. Datan, M. Höhse, A. Mantalaris, A. Bismarck, Towards a methodology for the effective surface modification of porous polymer scaffolds, Biomaterials 26 (36) (2005) 7537–7547.
[113] Y. Wan, C. Tu, J. Yang, J. Bei, S. Wang, Influences of ammonia plasma treatment on modifying depth and degradation of poly (L-lactide) scaffolds, Biomaterials 27 (13) (2006) 2699–2704.
[114] M. Domingos, F. Intranuovo, A. Gloria, R. Gristina, L. Ambrosio, P.J. Bártolo, P. Favia, Improved osteoblast cell affinity on plasma-modified 3-D extruded PCL scaffolds, Acta Biomater. 9 (4) (2013) 5997–6005.

Further reading

[115] F.G. Mangano, A. Caprioglio, L. Levrini, D. Farronato, P.A. Zecca, C. Mangano, Immediate loading of mandibular overdentures supported by one-piece, direct metal laser sintering mini-implants: a short-term prospective clinical study, J. Periodontol. 86 (2) (2015) 192–200.

Mechanical behavior of selective laser melting-produced metallic biomaterials

5

Wan Sharuzi Wan Harun, Kumaran Kadirgama*, Mahendran Samykano*, Devarajan Ramasamy*, Iswadi Ahmad†, Mahmoud Moradi‡*
*Faculty of Mechanical & Manufacturing Engineering, Universiti Malaysia Pahang, Pekan, Malaysia, †Kolej Kemahiran Tinggi MARA Kuantan, Kuantan, Malaysia, ‡Department of Mechanical Engineering, Malayer University, Malayer, Iran

5.1 Introduction

Biomaterials are substances used to repair damaged body parts through interaction with living systems. Advanced manufacturing techniques for biomaterials processing are continually being explored to minimize inventory, reduce cost, and achieve desired material properties. Selective laser melting (SLM) is currently the most popular powder bed fusion technique for manufacturing biomaterial metal products, as compared to other metal additive manufacturing (AM) methods [1–3]. Studies of SLM for metallic biomaterials are numerous and new research continues to appear [2, 4] as the technology matures, due to the advantages of SLM, such as high build resolution, excellent dimensional accuracy, clean build environment, low excess materials, and near-net-shape capability. The growth of SLM systems is related to new machinery developments and innovations to make the technology more productive and economically attractive. The new technology development and well-developed innovation for machine utilization became significant as the growth reached the year 2012 [5]. It included the in-house production of SLM systems for research involving metallic biomaterials such as cobalt-based alloys and nickel-titanium.

One of the unique things about the SLM process is its capability to manufacture complex geometrical structure instead of just simple, fully dense parts [6]. The structural design of the part is believed to influence the biomaterials in meeting specific mechanical characterizations directly [7, 8]. Therefore a great deal of research has explored the remarkable potential of SLM in manufacturing of difficult porous biomaterial parts. Ti6Al4V has become the most in-demand biomaterial used in porous structure fabrication using the SLM process.

This chapter reviews the roles of the SLM process in manufacturing various metals from the biomaterials family to develop their mechanical behavior. The review begins by introducing SLM mechanisms and addressing the processing parameters that influence the mechanical properties of metallic biomaterial parts. Then, the following discussion focuses on three different categories of SLM parts: fully dense (solid), porous,

Mechanical Behavior of Biomaterials. https://doi.org/10.1016/B978-0-08-102174-3.00005-X

and hybrid-based structures. Subsequently, the subsections discuss the mechanical behaviors of SLM parts for various metallic biomaterials and present some examples of SLMed metallic biomaterials applications. Finally, the concluding remarks underpin the current studies.

5.2 Selective laser melting

5.2.1 Mechanism

Commercially, the SLM process is rigid in terms of input processing, which includes the characteristics of the metal powder to be used. The general rule for the metals to be successfully processed in SLM is that the material must also be suitable for casting and welding processes. Therefore the commercially available raw metal powders are supplied by the additive machinery manufacturers, such as SLM Solutions. The set of widely available metallic biomaterials processible with SLM includes pure titanium (Ti), titanium alloys (Ti alloys), 316L stainless steel (316L SS), and cobalt-based alloys.

Fig. 5.1 shows a schematic drawing of the SLM process. The fabrication steps are similar to the binder jetting process. Briefly, the thermal source is employed to induce fusion between powder particles in a specific region of a build area, via layer-by-layer deposition, to produce a three-dimensional structure. Nevertheless, SLM processing has some notable differences compared to binder jetting. The powder is initially preheated until it reaches a suitable elevated temperature below its melting point. This preheating process is necessary to improve laser absorption, decrease the temperature gradient, minimize the thermal distortion of the final build structure, and improve wetting properties. Furthermore, SLM requires an oxygen-free environment to prevent

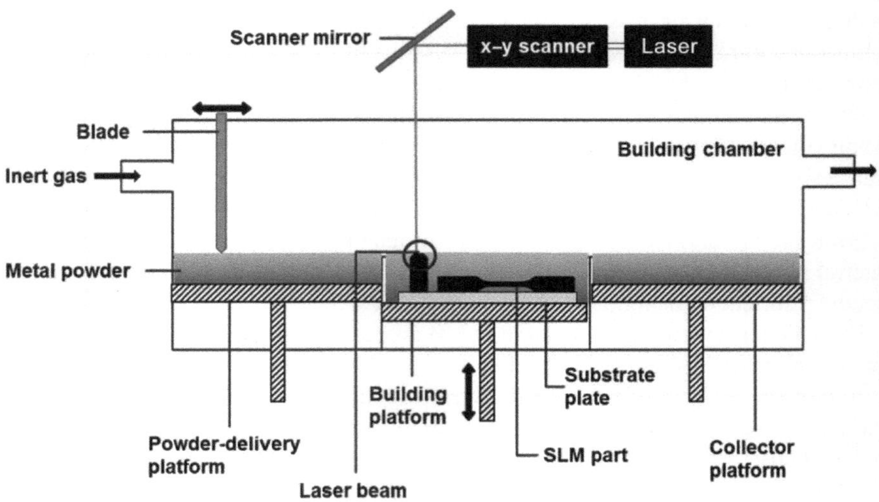

Fig. 5.1 SLM schematic diagram.

oxidation of the fabricated molten metal. Therefore, SLM usually operates in an inert gas environment, such as argon for reactive powder and nitrogen for nonreactive powder.

The SLM machine for powder handling consists of a platform to build the part from powder, to deliver the powder, and to collect the powder. It also has a blade (recoater) used to spread powder, a laser unit, and a scan head. First, the 3D model using the industrial standard stereo lithography (STL) file format is imported and sliced by the processing software to produce the laser scanning paths. The powder-delivery platform is then lifted up according to the powder height of the preset layer thickness from the building platform. The blade spreads the metal powder on a substrate plate attached to the building platform and forms the first layer before being melted by the laser. Once the first layer is completed, the building and collector platforms are lowered, while the powder-delivery platform is lifted up. For the first cycle of part fabrication, the powder-delivery platform is raised to a preset height, approximately the layer thickness of the parts. Subsequently, a new powder layer is created through the powder-delivery platform onto the substrate plate. The process of layer-by-layer melting of metal powder is repeated until the entire product is built. The remaining unfused or loose powder is removed from the machine and stored for the next process cycle.

5.2.2 Processing parameters

The SLM process involves many processing variables. The parameters used have been shown in the literature to influence the mechanical properties of the fabricated metallic biomaterial components. These parameters directly affect the thermal cycle experienced by the powder metals during fabrication and consequently influence the microstructure and mechanical properties of both static and cyclic loadings.

Various processing parameters are involved in the SLM process, including laser-dependent parameters as well as build-dependent parameters. The process variables of the SLM process can be organized as shown in Fig. 5.2.

It is essential to understand the interaction between these parameters. Different combinations of the processing parameters alter the integrity of the biomaterials processed. Therefore, many efforts have been made to understand the correlations between these parameters and how they influence the mechanical properties of the fabricated components. Thus identification of the optimal manufacturing is required to manufacture SLM products with the desired final mechanical properties.

Research shows that scanning speed and laser power can have significant effects on the structure of the final product. These parameters determine the energy transferred to the metal powders, which directly affects the dependent melting temperature for the liquid phase in the melted components. An article by Attar et al. demonstrated the combination of these parameters on commercially pure titanium and how it affected the mechanical properties of the fabricated parts [9]. The melting mechanism of the combined processing parameters is summarized in Table 5.1.

Layer thickness is also known to influence the porosity of the fabricated parts. The porosity then influences the mechanical properties of the fabricated part, such as the tensile strength and hardness. Moreover, as the layer thickness decreases, the

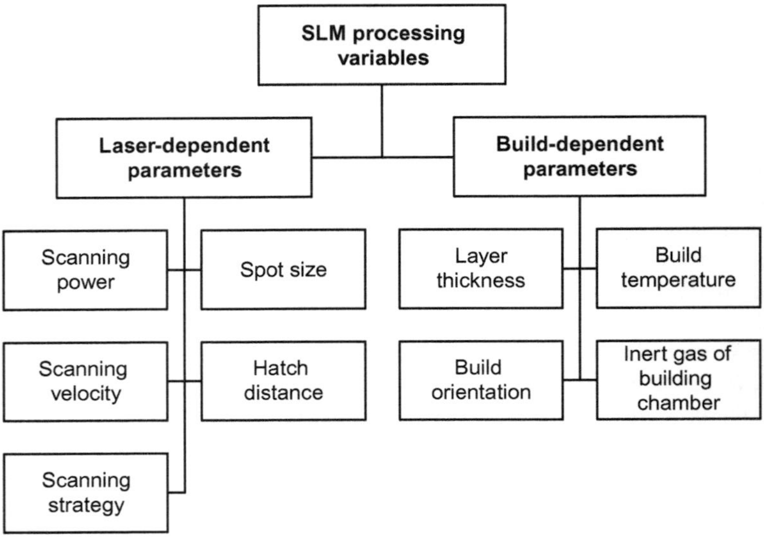

Fig. 5.2 Common SLM processing parameters that influence the manufacturing process.

Table 5.1 Conditions of melting mechanisms with combined processing parameters

Condition	Description
(a) No melting	Delivered laser energy is inadequate to affect the powder and left a lot of unfused powder particles.
(b) Partial melting	Combination of low scan speed (<0.06 m/s) and medium laser power formed a liquid phase on the surface of particles. After crystallization, unmelt cores of particles joined into the coarsened balls. The surface illustrated the first type of balling phenomena [10].
(c) Melting with balling	Combination of high scan speed (≥0.06 m/s) and high laser power generated thin, long, and cylindrical lines of liquid scan track shapes, which later broke up into rows of coarsened beads. This shows the results of surface tension reduction.
(d) Complete melting	Enough laser energy formed permanent tracks of molten metal material. A fully melted compact solid surface formed continuous lines by the tracks formation.

obtainable porosity also decreases, when other processing parameters are controlled. Porosity is less significant for smaller layer thickness. It is known that, in a loose powder, a considerable amount of air is present in the space between particles. When the layer of loose powder experiences a laser-induced melting, the existed space or void tends to become trapped bubbles [11]. In a condition where the laser energy input is fixed and the layer is thinner, the laser beam can penetrate deeper into the layer, hence leading to a higher degree of melting. When this happens, most of the trapped air bubbles, including those at the bottom of the layer, can rise to the layer's surface and break as the liquid quickly solidifies, thus leading to the formation of porosity.

Another vital parameter is building orientation, Zwhich is defined as the acute angle between the vertical axis of the build platform and the longitudinal axis of a given product. Significant anisotropy effects on the building orientation can be observed, as recorded in research done by Shifeng et al. [12]. The tensile property of fabricated 316L stainless steel parts was found to show better ductility when horizontally built, while the vertically built parts showed an optimal combination of ductility and strength. This phenomenon can be interpreted as being related to the elongated grain structure orientation. The grain orientation is dependent on the solidification phase conditions. In general, the grain growth direction is from the cooler to the warmer region. In other words, the top surface is exposed to the laser beam, whereas the bottom surface lies on a solid substrate. The orientation of the elongated grain is in the direction of the thermal gradient. The melting lines lie in a plane that is aligned with the build platform. Hence, the fabricated parts built at different orientations have different elongated grain orientations, and the tensile properties are also different.

5.3 Categories of SLMed products

The mature SLM technology is capable of manufacturing various geometrical structures. Depending on the applications, a variety of metallic biomaterials are processed with different final product designs. The most common metallic biomaterial products processed by SLM can be categorized into three groups, as shown in Fig. 5.3.

5.3.1 Solid parts

The SLM process can manufacture a solid part with a high relative density of up to 99% with the appropriate manipulation of scanning parameters and scanning strategy.

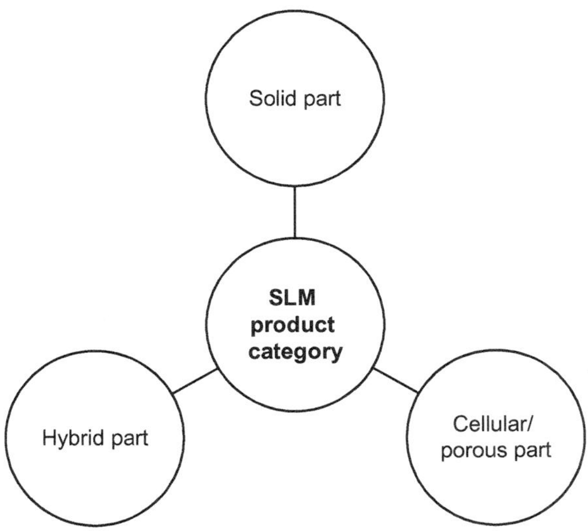

Fig. 5.3 Categories of SLMed fabricated products.

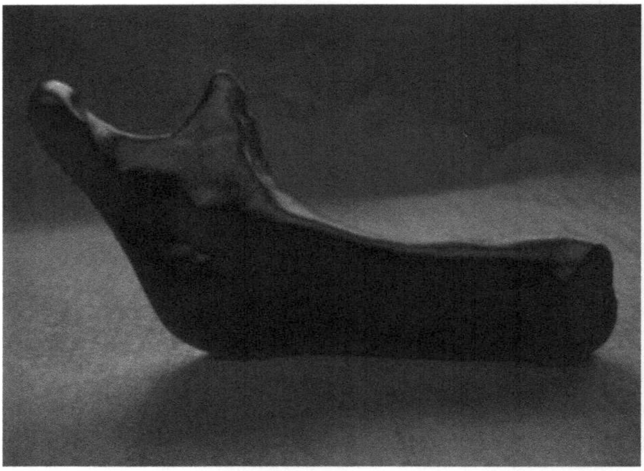

Fig. 5.4 Solid Ti6Al4V mandible implant fabricated by SLM process [13].

Although the parts with solid forms are exposed to defects such as porosity, cracks, or buckling, a proper selection of the processing strategy can help to minimize these drawbacks. Optimal scanning parameters and scanning strategies are used to allow better fusion bonding between layers during fabrication, reduction of possible porosity, prevention of surface swelling, and minimization of residual stresses. Several methods have been used to improve the part's density such as laser remelting, application of the semiempirical formula for power energy density utilization, laser power increment, and others. Fig. 5.4 shows a solid Ti6Al4V mandible implant fabricated through the SLM process.

Meanwhile, SLM is believed to help improve the mechanical properties of the products made from metallic biomaterials. This is proven in the research work reported by Bartolomeu et al. in Table 5.2 [14]. Higher values of hardness, yield strength, and ultimate tensile strength and the lower ductility values of the fabricated parts were improvements, as compared to the parts built by hot pressing (HP) and conventional casting methods. This is explained further in Section 5.4.4.

5.3.2 Porous/cellular parts

SLM is also a promising technique for the manufacturing of net-shape parts using various metals with porous, cellular, or lattice structures [15, 16]. The lattice structure

Table 5.2 Mechanical properties of 316L stainless steel solid parts fabricated by SLM, HP, and conventional casting [14]

Condition	Hardness (HV)	Yield strength (MPa)	Ultimate tensile strength (MPa)	Elongation (%)
SLM	~229	~490	~630	~25
Hot press	~176	~270	~570	~32
Casting	~165	~200	~450	~45

is a unique type of porous structure. It consists of an ordered network structure that connects groups of the unit cell. The unit cell is usually built from struts or beams that create a 3D structure. Different types of commonly used unit cells include cubic, diamond, truncated cuboctahedron, and some other complex types, such as plane and sheet gyroid. Metals from the biomaterial family, such as Ti6Al4V and CoCr, are considered to be porous metallic biomaterials when used to make these structures.

The characteristics of a porous biomaterial product are determined by its parent materials as well as the properties of the unit cell set used to form the structure, including type, size, and relative density. The relative density of the porous part is defined as a measure of a fraction of a solid region in the porous structure. Hence, it can be said that the mechanical properties of the porous biomaterial part can be modified by altering the parent material and the properties of its unit cell. Fig. 5.5 shows an example of fabricated porous Ti6Al4V ELI [17].

5.3.3 Hybrid parts

Advances in SLM have allowed the capability to fabricate advanced complex structures of a hybrid form, where solid matrix and controlled porous structures are manufactured to be a single component: e.g., the structures for the acetabular cup, hip stem, or other orthopedic implants, in which the surface geometry of the fabricated component helps to enhance osseointegration while the internal section is fully dense.

Fig. 5.5 Porous structure of metallic biomaterial Ti6Al4V ELI constructed from diamond-shaped unit cells [17].

Functionally gradient materials (FGMs) are materials with a specific type of structural design that falls under the hybrid group. In FGMs, the materials are designed with continuous changes or stepwise changes in their properties and microstructures. They allow material tailoring with a particular structure, morphology, and mechanical properties. The FGMs play a role in stress-shielding effect reduction of femoral stem implants and bone. Stiffness configuration strategies can be used to match the stiffness between the bone and the stem.

Porosity is introduced into a design, as it can alter the mechanical properties of a component, hence highlighting the potential of single alloy manufacturing for the functionally graded component. A lot of research has been carried out on applying FGMs in SLM CoCrMo femoral stem fabrication [18]. The studies utilized porous cellular structures with a unit cell approach as their stiffness configuration strategy. Table 5.3 shows how the variation of strut and pore sizes with a range between 0.50–1.50 and 1.57–1.88 mm, respectively, can lead to different effective elastic modulus, effective compressive yield strength, and stiffness of the fabricated CoCrMo femoral stems. The stiffness of the functionally graded design and the yield strength of the fully dense CoCrMo with 1 mm strut size of the cellular structure area were found to be 600 and 175 MPa, respectively.

Other research was done on 316L stainless steel to produce a part with a pore gradient structure [19]. Scan speed (processing parameter) was changed in a gradient manner to produce a part with a pore gradient structure. In this configuration, different scan speeds varying from 90 mm/s up to 180 mm/s led to variations in

Table 5.3 Recent functionally gradient materials research works utilizing SLM

Processing or design configuration		**Mechanical properties**			**Ref.**
Strut size (mm)	**Pore size (mm)**	**Effective elastic modulus (GPa)**	**Effective compressive yield strength (MPa)**	**Mean stiffness (N/mm)**	**[18]**
CoCrMo femoral stem					
1.50	1.88	17.98	295.72	394.45	
1.00	1.80	13.64	175.15	351.57	
0.50	1.57	4.79	65.43	309.32	

Processing or design configuration	**Mechanical properties**		**Ref.**
Scan speed (mm/s)	**Porosity (%)**	**Ultimate tensile strength (MPa)**	**[19]**
316L stainless steel			
90	4	650	
120	11	410	
150	22	120	
180	35	30	

powder melting conditions during the SLM process, causing the formation of higher porosity that aligned with the increase in scan speed, as shown in Table 5.3. This directly affected the mechanical property, i.e., the ultimate tensile strength of the 316L stainless steel part.

5.4 Mechanical characterizations

A summary of mechanical properties for commonly fabricated solid biomaterial parts using SLM technology is presented in Table 5.4. The fabricated products were made from 316L stainless steel, commercially pure titanium (CP-Ti), Ti6Al4V, cobalt chromium molybdenum (CoCrMo), AlSi10Mg, nickel-titanium, and tantalum.

5.4.1 Tensile and elongation

Static tensile strength of the SLM parts depends on the density of the parts as well as the microstructures formed during the process. The microstructures of SLM parts are finer than with other classical fabrication routes, such as casting. This will be explained further in Section 5.4.4. The yield strength, ultimate tensile strength, and elongation of SLM fabricated biomaterials are summarized in Table 5.4. The highlighted values show that the tensile and elongation properties of the SLM fabricated biomaterials are comparable or better than those of bulk biomaterials. The performance of SLM fabricated biomaterials is influenced by the processing parameters, such as scanning power and building orientation.

The increase in both yield and ultimate tensile strength of the products is due to grain refinement, which is defined by the Hall-Petch strengthening. With regard to ductility, the higher ductile failure mode, or better elongation amount, of the SLM

Table 5.4 Summary of mechanical properties of different types of metallic biomaterials processed by SLM method

Materials	Vickers hardness (HV)	Yield strength (MPa)	Ultimate tensile strength (MPa)	Elongation (%)	Fatigue limit (MPa)	Ref.
316L stainless steel	210–240	430–510	510–620	30–75	265–365	[14, 20, 21]
Ti6Al4V	~406	~1000	~1100	~8	N/A	[22, 23]
CP-Ti	~261	~555	~757	~19	N/A	[9]
CoCrMo	410–515	428–873	1159–1303	9–12	389–492	[24, 25]
AlSi10Mg	~1196	~241	~384	~6	125	[26]
Nickel-titanium	540–735	N/A	N/A	N/A	N/A	[27]
Tantalum	~425	~450	~739	~2	N/A	[28]

parts compared to wrought biomaterials is due to the lower residual porosity of ~0.1%. However, a significant elongation reduction can be observed if there is a high residual porosity of 2.4%, which subsequently leads to brittle failure modes [29].

It is obvious that the tensile strength depends on the microstructures, which also exhibit anisotropic features. For instance, the wide ranges of yield strength, ultimate tensile strength, and elongation shown by the 316L stainless steel in Table 5.4 are contributed by the building orientation. Parts built at 45 degrees are normally reported as having higher yield strength and ultimate tensile strength than those of 90-degree builds [20], while the elongation of parts built at 45 degrees is shown to be reduced by up to 50% as compared to that of 90-degree builds.

5.4.2 Hardness

The hardness of SLM parts is influenced by density and microstructure. The parts produced are usually high in density due to minimum porosity, which leads to an improvement in the hardness value of the parts. Table 5.4 shows that the hardness value of SLM-produced biomaterials is higher compared to that of the wrought process. An example of hardness difference for hot pressing and conventional casting of 316L stainless steel can be seen from the values of 176 and 165 HV, respectively [14]. The values are much lower compared to the hardness values of 210–240 HV obtained from SLM fabrication.

Another reason that hardness of SLM parts is higher than that of wrought-fabricated parts is the condition of the microstructure phases. For example, an SLM Ti6Al4V part was found to have higher hardness values than wrought due to the presence of martensite phase regimes and grain refinement. This can also be seen by the hardness value of SLM-processed CP-Ti at 261 HV (shown in Table 5.4), which is higher than that of the casting-produced part of 210 HV.

5.4.3 Fatigue

Fatigue life is significantly moderate compared to wrought products. This is due to various factors such as microstructure, porosity, surface finish, and residual stress [30]. Nevertheless, properties inherited from the SLM process, such as surface roughness and defects, significantly influence the fatigue performance of the parts. Similar to the layer-wise manufacturing process, the SLM process causes an increase in surface roughness due to the stair-stepping effect, which leads to stress concentration and early failure of the as-built parts under fatigue loading. Material defects such as porosity and inadequate layer bonding cause the experimental data to scatter. Therefore postprocessing treatments such as mechanical surface treatment (i.e., polishing), which cures and densifies the biomaterials via hot-isostatic pressing, help to improve the fatigue properties to be comparable to the existing wrought and as-cast parts from metallic biomaterials.

As an example, research by Riemer et al. showed that as-built 316L stainless steel with an SLM surface was found to exhibit a fatigue limit of 108 MPa [31]. This was lower compared to hot isostatically pressed with turned surface parts and traditionally processed parts, with values of 317 and 240–381 MPa, respectively. In order to reach the fatigue limits of 265–365 MPa for the 316L stainless sample, as reported in Table 5.4, the as-built SLM needs to undergo either surface modification or heat treatment to reduce any residual stresses and to allow an increase in the fatigue limit value.

5.4.4 Microstructure

Better mechanical behaviors are exhibited by the SLM parts compared to other fabrication technologies, especially concerning hardness and tensile properties due to the grain size decrease. Smaller grain size requires higher energy for a dislocation to alter its direction and move into the adjacent grain when there are a considerable number of grain boundaries to shift forward. This situation is related to the SLM process condition. From the analysis of microstructure in Fig. 5.6, the formation of the fine microstructure is responsible for the higher hardness and strength attained by the SLM parts.

The finer grains of the SLM parts compared to the molded or as-cast part morphology are believed to be due to the rapid solidification of laser-heated metallic biomaterials. The heat is rapidly conducted from the molten zone into the surrounding metal. The rapid cooling can initiate the formation of unbalanced crystal phases as shown by the example in Fig. 5.7 [32]. Fig. 5.7 shows the metallographic of SLMed cobalt chrome fabricated using recommended parameters by the machine manufacturer. Fig. 5.7C illustrates the grains with approximate sizes of 0.3–0.6 μm. During

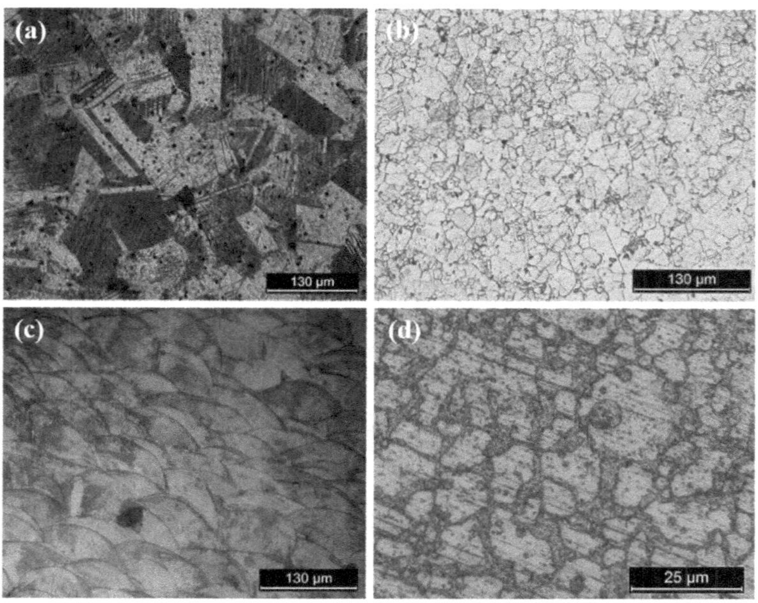

Fig. 5.6 Optical micrograph images of 316L stainless steel parts produced by: (A) casting, (B) HP, (C) and (D) SLM [14].

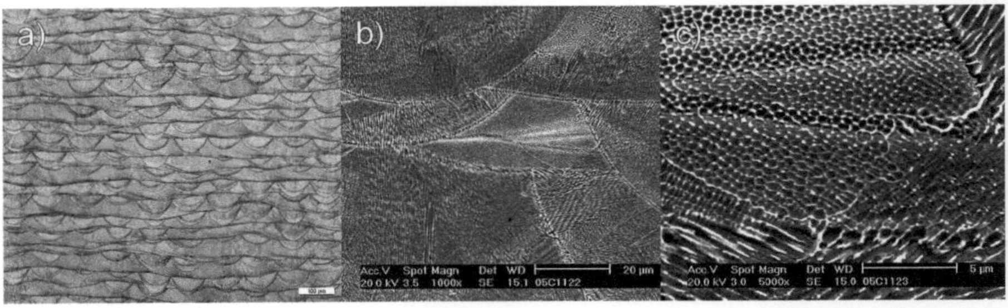

Fig. 5.7 SLM cobalt chrome at (A) 10×, (B) 1000×, and (C) 5000× magnification [32]. From M. Shellabear, O. Nyrhilä, Advances in materials and properties of direct metal laser-sintered parts. Available from https://www.rm-platform.com/downloads2/send/5-papers/255-paper-lane-2007-eos-shellabear.

the layer-by-layer fabrication, individual lines of biomaterial were slowly melted and recrystallized. Hence, it was expected that the majority of SLM parts would exhibit a fine-grained structure similar to that shown in Fig. 5.7.

5.5 SLM applications

As one of the "nonclassic" manufacturing complex part methods under the group of additive manufacturing processes, the SLM technique has been approved by the biomedical and aerospace industries. To align with the current medical field trend towards individuality, SLM is expected to fabricate custom, complex, and more flexible design parts, for example, an implant for an individual patient. Lattice structure designs, for instance, are very suitable for orthopedic implants or bone substitutes. For these applications, biomaterials, or specifically novel porous biomaterials with similar mechanical properties to bone, are preferred, to help minimize the stress-shielding effects. Moreover, the porous structures of such biomaterials provide a substantial interconnected volume of bone tissue ingrowth. The aerospace industry can also find benefits from these structures, as they provide structural components with enhanced strength-to-weight and stiffness-to-weight ratios.

Table 5.5 shows examples of SLM utilization in the orthopedics field. The earliest application was in 2005 in the fabrication of the lower jaw structure, lumbar vertebra, and tubular femoral bone, as shown in Table 5.5 (a) [33]. The MCP Realizer SLM machine was used to fabricate the 316L stainless steel lower jaw structure and lumbar vertebra, and a pure titanium tubular femoral bone. Other SLM-fabricated implants include customized Ti6Al4V ELI frontoparietal-temporal bone that fits precisely the patient's cranial defect for craniomaxillofacial surgery (Table 5.5 (b)) [39]; and a talar component for CoCrMo endoprosthetic ankle devices (Table 5.5 (c)). The latter component underwent a kinematic test carried out on a cadaver leg. It was revealed that the tailor-fit articular surfaces reproduced the natural joint motion well, thus supporting the suitability of SLM for patient personalization in total ankle replacement. SLM techniques have also been studied for processing metallic biomaterials in other

Table 5.5 List of implants and devices fabricated by SLM processes

Materials	Fabricated implants and devices	Ref.
316L stainless steel	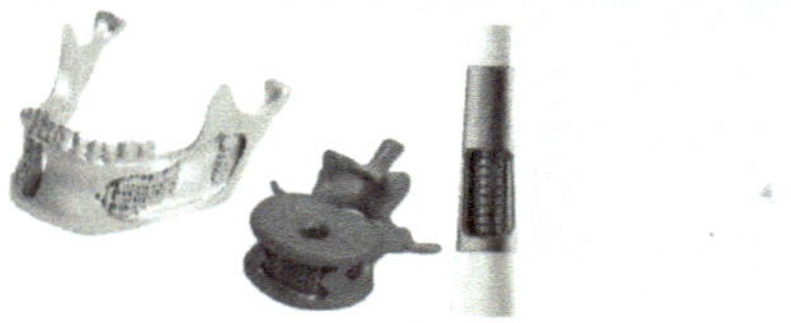 (a) (From left) lower jaw, lumbar vertebra, and tubular femoral bone [33]	[13, 33–38]
Ti6Al4V ELI	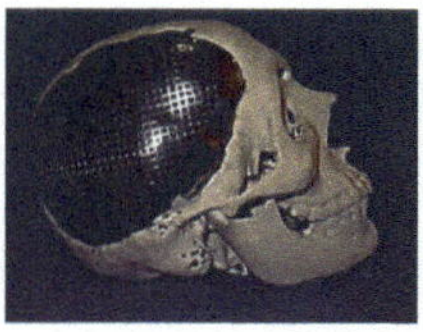 (b) Frontoparietal-temporal bone [39]	
CoCrMo	 (c) Talar component [40]	

applications such as oral screw-retained implants, hip endoprostheses, acetabular cups, and other biomedical devices [13, 34–37].

5.6 Conclusions

SLM allows the fabrication of fully dense parts and exceptional complex geometrical structures. Although SLM is rigid in terms of input processing, which includes the characteristics of the metal powder to be used, it can fabricate various metallic biomaterials such as 316L stainless steel, commercially pure titanium (CP-Ti), Ti6Al4V, cobalt chromium molybdenum (CoCrMo), AlSi10Mg, nickel-titanium, and tantalum. Processing parameters and the various SLM manufacturing designs of solid, porous, and hybrid products influence the final fabricated mechanical properties of SLM products. The SLM method is used to improve the mechanical properties of metallic biomaterial parts as compared to wrought and as-cast parts and to achieve the desired mechanical properties. With the improved mechanical behavior of metallic biomaterials, SLM has been adopted in the biomedical and aerospace industries.

Acknowledgments

The authors gratefully acknowledge the support from Universiti Malaysia Pahang, Qatar National Research Fund NPRP88762375 (UIC161504), Universiti Malaysia Pahang internal grants RDU140354, RDU160337, RDU150337, RDU160354, GRS1503115, GRS1503145, PGRS170388, and the Research Acculturation Grant Scheme (RAGS) and Race Acculturation Collaborative Efforts (RACE) provided by the Ministry of Higher Education, Malaysia RDU151404 and RDU151314, respectively.

References

[1] W. Harun, R. Asri, J. Alias, F. Zulkifli, K. Kadirgama, S. Ghani, et al., A comprehensive review of hydroxyapatite-based coatings adhesion on metallic biomaterials, Ceram. Int. (2017).
[2] S. Ghani, M. Zakaria, W. Harun, Z. Zaulkafilai, in: Dimensional accuracy of internal cooling channel made by selective laser melting (SLM) and direct metal laser sintering (DMLS) processes in fabrication of internally cooled cutting tools, MATEC Web of Conferences, 2017, p. 01058.
[3] S.A.C. Ghani, S.R. Mohamed, W.S.W. Harun, N.A.Z.M. Noar, in: Physical evaluations of Co-Cr-Mo parts processed using different additive manufacturing techniques, AIP Conference Proceedings, 2017, p. 100001.
[4] R.I.M. Asri, W.S.W. Harun, M. Samykano, N.A.C. Lah, S.A.C. Ghani, F. Tarlochan, et al., Corrosion and surface modification on biocompatible metals: a review, Mater. Sci. Eng. C 77 (2017) 1261–1274.
[5] W.S.W. Harun, M.S.I.N. Kamariah, N. Muhamad, S.A.C. Ghani, F. Ahmad, Z. Mohamed, A review of powder additive manufacturing processes for metallic biomaterials, Powder Tech. 327 (2018) 128–151.
[6] N.K.M.S. Islam, W.S.W. Harun, S.A.C. Ghani, Z. Ghazalli, D.N.A. Shri, M.A.H.M. Adib, et al., Manufacture by selective laser melting & physical behaviour of commercial 316L stainless steel, in: International Medical Device and Technology Conference 2017, (2017), pp. 87–89.
[7] N.K.M.S. Islam, W.S.W. Harun, S.A.C. Ghani, M.A. Omar, M.H. Ramli, M.H. Ismail, in: Physical properties and microstructure study of stainless steel 316L alloy fabricated by selective laser melting, AIP Conference Proceedings, 2017, p. 100015.
[8] M. Kamariah, W. Harun, N. Khalil, F. Ahmad, M. Ismail, S. Sharif, in: Effect of heat treatment on mechanical properties and microstructure of selective laser melting 316L stainless steel, IOP Conference Series: Materials Science and Engineering, 2017, p. 012021.
[9] H. Attar, M. Calin, L.C. Zhang, S. Scudino, J. Eckert, Manufacture by selective laser melting and mechanical behavior of commercially pure titanium, Mater. Sci. Eng. A 593 (2014) 170–177.
[10] D. Gu, Y. Shen, Processing conditions and microstructural features of porous 316L stainless steel components by DMLS, Appl. Surf. Sci. 255 (2008) 1880–1887.
[11] A.N. Chatterjee, S. Kumar, P. Saha, P.K. Mishra, A.R. Choudhury, An experimental design approach to selective laser sintering of low carbon steel, J. Mater. Process. Technol. 136 (2003) 151–157.
[12] W. Shifeng, L. Shuai, W. Qingsong, C. Yan, Z. Sheng, S. Yusheng, Effect of molten pool boundaries on the mechanical properties of selective laser melting parts, J. Mater. Process. Technol. 214 (2014) 2660–2667.

[13] L.S. Bertol, W.K. Júnior, F.P. da Silva, C. Aumund-Kopp, Medical design: direct metal laser sintering of Ti–6Al–4V, Mater. Des. 31 (2010) 3982–3988.
[14] F. Bartolomeu, M. Buciumeanu, E. Pinto, N. Alves, O. Carvalho, F.S. Silva, et al., 316L stainless steel mechanical and tribological behavior—a comparison between selective laser melting, hot pressing and conventional casting, Addit. Manuf. 16 (2017) 81–89.
[15] N.A. Malek, S. Mohamed, S.C. Ghani, W.W. Harun, in: Critical evaluation on structural stiffness of porous cellular structure of cobalt chromium alloy, IOP Conference Series: Materials Science and Engineering, 2015, p. 012019.
[16] Z.A.M. Taib, W.S.W. Harun, S.A.C. Ghani, M.F.F.A. Rashid, M.A. Omar, H. Ramli, Dimensional accuracy study of open cellular structure CoCrMo alloy fabricated by selective laser melting process, Adv. Mater. Res. 1133 (2016) 280.
[17] B. Van Hooreweder, Y. Apers, K. Lietaert, J.-P. Kruth, Improving the fatigue performance of porous metallic biomaterials produced by Selective Laser Melting, Acta Biomater. 47 (2017) 193–202.
[18] K.B. Hazlehurst, C.J. Wang, M. Stanford, An investigation into the flexural characteristics of functionally graded cobalt chrome femoral stems manufactured using selective laser melting, Mater. Des. 60 (2014) 177–183.
[19] R. Li, J. Liu, Y. Shi, M. Du, Z. Xie, 316L stainless steel with gradient porosity fabricated by selective laser melting, J. Mater. Eng. Perform. 19 (2010) 666–671.
[20] E. Liverani, S. Toschi, L. Ceschini, A. Fortunato, Effect of selective laser melting (SLM) process parameters on microstructure and mechanical properties of 316L austenitic stainless steel, J. Mater. Process. Technol. 249 (2017) 255–263.
[21] R. Casati, J. Lemke, M. Vedani, Microstructure and fracture behavior of 316l austenitic stainless steel produced by selective laser melting, J. Mater. Sci. Technol. 32 (2016) 738–744.
[22] N. Guo, M.C. Leu, Additive manufacturing: technology, applications and research needs, Front. Mech. Eng. 8 (2013) 215–243.
[23] C. Lancea, L.-A. Chicos, S.M. Zaharia, M.A. Pop, in: Microstructure and micro-hardness analyses of titanium alloy Ti-6Al-4V parts manufactured by selective laser melting, MATEC Web of Conferences, 2017, p. 03009.
[24] B. Qian, K. Saeidi, L. Kvetková, F. Lofaj, C. Xiao, Z. Shen, Defects-tolerant Co-Cr-Mo dental alloys prepared by selective laser melting, Dent. Mater. 31 (2015) 1435–1444.
[25] Y. Kajima, A. Takaichi, T. Nakamoto, T. Kimura, Y. Yogo, M. Ashida, et al., Fatigue strength of Co–Cr–Mo alloy clasps prepared by selective laser melting, J. Mech. Behav. Biomed. Mater. 59 (2016) 446–458.
[26] N.E. Uzan, R. Shneck, O. Yeheskel, N. Frage, Fatigue of AlSi10Mg specimens fabricated by additive manufacturing selective laser melting (AM-SLM), Mater. Sci. Eng. A 704 (2017) 229–237.
[27] M. Elahinia, N. Shayesteh Moghaddam, M. Taheri Andani, A. Amerinatanzi, B.A. Bimber, R.F. Hamilton, Fabrication of NiTi through additive manufacturing: a review, Prog. Mater. Sci. 83 (2016) 630–663.
[28] L. Zhou, T. Yuan, R. Li, J. Tang, G. Wang, K. Guo, Selective laser melting of pure tantalum: Densification, microstructure and mechanical behaviors, Mater. Sci. Eng. A 707 (2017) 443–451.
[29] D. Herzog, V. Seyda, E. Wycisk, C. Emmelmann, Additive manufacturing of metals, Acta Mater. 117 (2016) 371–392.
[30] P. Edwards, M. Ramulu, Fatigue performance evaluation of selective laser melted Ti–6Al–4V, Mater. Sci. Eng. A 598 (2014) 327–337.

[31] A. Riemer, S. Leuders, M. Thöne, H.A. Richard, T. Tröster, T. Niendorf, On the fatigue crack growth behavior in 316L stainless steel manufactured by selective laser melting, Eng. Fract. Mech. 120 (2014) 15–25.

[32] P. Hanzl, M. Zetek, T. Bakša, T. Kroupa, The Influence of Processing Parameters on the Mechanical Properties of SLM Parts, Procedia Eng. 100 (2015) 1405–1413.

[33] M. Wehmöller, P.H. Warnke, C. Zilian, H. Eufinger, Implant design and production—a new approach by selective laser melting, Int. Congr. Ser. 1281 (2005) 690–695.

[34] B. Vandenbroucke, J.P. Kruth, Selective laser melting of biocompatible metals for rapid manufacturing of medical parts, Rapid Prototyp. J. 13 (2007) 196–203.

[35] T. Habijan, C. Haberland, H. Meier, J. Frenzel, J. Wittsiepe, C. Wuwer, et al., The biocompatibility of dense and porous Nickel–Titanium produced by selective laser melting, Mater. Sci. Eng. C 33 (2013) 419–426.

[36] J. Liu, G. Han, S. Pan, Y. Ge, H. Feng, Z. Shen, Biomineralization stimulated peri-titanium implants prepared by selective laser melting, J. Materiomics 1 (2015) 253–261.

[37] A.L. Jardini, M.A. Larosa, M.F. Macedo, L.F. Bernardes, C.S. Lambert, C.A.C. Zavaglia, et al., Improvement in cranioplasty: advanced prosthesis biomanufacturing, Procedia CIRP 49 (2016) 203–208.

[38] D.A. Hollander, M. von Walter, T. Wirtz, R. Sellei, B. Schmidt-Rohlfing, O. Paar, et al., Structural, mechanical and in vitro characterization of individually structured Ti–6Al–4V produced by direct laser forming, Biomaterials 27 (2006) 955–963.

[39] A.L. Jardini, M.A. Larosa, R.M. Filho, C.A.D.C. Zavaglia, L.F. Bernardes, C.S. Lambert, et al., Cranial reconstruction: 3D biomodel and custom-built implant created using additive manufacturing, J. Craniomaxillofac. Surg. 42 (2014) 1877–1884.

[40] E. Liverani, A. Fortunato, A. Leardini, C. Belvedere, S. Siegler, L. Ceschini, et al., Fabrication of Co–Cr–Mo endoprosthetic ankle devices by means of Selective Laser Melting (SLM), Mater. Des. 106 (2016) 60–68.

Further reading

[41] M. Shellabear, O. Nyrhilä, Advances in materials and properties of direct metal laser-sintered parts. Available from https://www.rm-platform.com/downloads2/send/5-papers/255-paper-lane-2007-eos-shellabear.

Machining of a biomaterial with dual negative tool geometry

6

C.H. Lauro*,‡, J.C. Nogueira*, D. Carou*,†, L.C. Brandão‡, J. Paulo Davim*
*Department of Mechanical Engineering, University of Aveiro, Aveiro, Portugal,
†Department of Mechanical and Mining Engineering, University of Jaén, Jaén, Spain,
‡Department of Mechanical Engineering, Centre for Innovation in Sustainable Manufacturing, Federal University of São João del Rei, São João del Rei, Brazil

6.1 Introduction

Interest in the manufacture of new materials has been growing over the last few years because of a greater need to manufacture enhanced parts and products of high quality. Machinability is an index that compares two materials, one being a reference material defined as easy to machine. This index is widely studied and it is useful in analyzing machining processes. According to Trent and Wright [1], machinability can be related to the productivity, costs, and quality of the finishing of a machined component. Accordingly, an improved knowledge of machining processes is important to obtain low-cost components, taking into account that the cost of machining can represent at least 15% of the cost of the product [2]. This topic has attracted the interest of researchers to study the behavior of materials with poor machinability. Examples of these materials are SAE XEV-F steel employed in automobile components [3], and AISI H13 and AISI D2 steels for forming tools [4, 5].

Titanium and its alloys are considered difficult-to-machine materials because of their high hardness, strength, and toughness resulting in elevated temperatures during machining [6]. Moreover, the heat dissipation in the machining of titanium alloys is hampered by the low thermal conductivity and high specific heat [6, 7]. The risk of fire must be taken into account when the temperature reaches values higher than 400°C [7]. Because of these drawbacks and the increasing number of applications in sectors such as aerospace, automotive, and biomedical [7], the study of titanium machining is of great importance.

Because of the problems associated with heat evacuation, the effect of different cooling/refrigeration alternatives in the machining of titanium alloys has been analyzed. For instance, the use of cryogenic cooling was analyzed in drilling of the Ti-6Al-4V alloy [8]. When compared to wet conditions, the authors observed that lower values of temperature were obtained with cryogenic cooling. However, the measured values of the cutting force and surface roughness were higher. Hui et al. [9] analyzed the electrical discharge machining (EDM) process. The cryogenic cooling of the electrode decreased the discharge gap, improving the machining accuracy.

Mechanical Behavior of Biomaterials. https://doi.org/10.1016/B978-0-08-102174-3.00006-1

Among the titanium alloys, most biomedical applications use the Ti-6Al-4V alloy. However, new alloys are being developed to replace it that eliminate the use of vanadium, since the presence of vanadium ions in human tissues may cause serious health problems [10].

According to Wu and Guo [11], the quality of machined parts made of titanium alloys can be improved by selecting adequate tool geometries. The authors investigated the influence of the clearance and rake angles in the turning of the Ti-6Al-2Zr-2Sn-3Mo-1Cr-2Nb alloy. When increasing both angles, the cutting forces decreased. Usually, medical components require low values of surface roughness because, generally, bone-forming cells attach themselves to surfaces with roughness in the nanometer range [12]. Gupta et al. [13] optimized the turning of commercial pure titanium, Grade II, with cubic boron nitride (CBN) tools and minimum quantity lubrication (MQL) cooling. The optimization model was developed employing artificial intelligence and statistical techniques that provided close results.

In the machining of materials having poor machinability, it is suggested that the rake angle should be negative. However, a negative rake angle presents some drawbacks, such as (a) compression of the tool material caused by the normal force; and (b) lower tool life due to the contact interface tool/chip with temperature shift toward the cutting edge [14]. Furthermore, the use of this geometry requires that the cutting setting be properly selected to avoid vibrations due to the increase in the cutting forces, which has a negative impact on the stability of the cutting [15]. Ozturk and Altan [16] observed that the cutting forces grew proportionally to the decrease of the rake angle and the uncut chip thickness, but they diminished when cutting speed was increased.

According to Fang [17], traditional machining theories were developed based on positive or slightly negative rake angles. The author highlighted that the tool-chip friction studies employing negative rake angle tools are still limited. However, in the last several years, some studies were developed to improve knowledge of the chip formation mechanism in machining processes such as hard turning, micromachining, and others.

Furthermore, to select the tool geometry, several variables should be analyzed, such as the inclination angle, which influences the chip direction flow, and the use of a chip breaker, which often dictates the resulting rake angle [14]. Sandvik Coromant [18] provides indications of specific characteristics such as negative inserts, clearance angle of 0 degrees, possibility of use of both sides, and high edge strength. The manufacturer suggests the use of the negative inserts for external turning and heavy cutting conditions. This insert is mounted in the tilted holder, i.e., holder with a negative inclination angle. The tool used to generate grooves provides chips of small size and shape, which facilitates handling and disposal, but its geometry complicates the prediction of tool wear and tool life [19]. The use of a chip breaker was also analyzed in experimental studies. For instance, in turning of the Ti-15V-3Al-3Cr-3Sn alloy, Maurotto et al. [20] employed a cemented-carbide tool with chip breaker, improving the results for the feed rate of 0.1 mm/rev.

This chapter analyzes the machinability of the Ti-6Al-7Nb titanium alloy employing tools with dual negative tool geometry, which is characterized by a chip breaker and negative inclination and rake angles. This geometry was employed because manufacturing of implants requires high surface quality in the machining, which can involve interrupted cutting in materials having poor machinability. Thus, the objective was to improve understanding of the effect of these tools, compared with theoretical values, on the surface roughness, and the relation between cutting parameters and the response of the tool/material pair.

6.2 Material and methods

The tests were developed using bars of Ti-6Al-7Nb alloy with hardness 32 HR_C, provided by TiFast S.r.l. The chemical composition of the alloy was: aluminum (5.94%–6.02%), carbon (0.013%), ferrous (0.15%–0.16%), hydrogen (0.003%), nitrogen (0.005%), niobium (6.83%–6.90%), oxygen (0.169%–0.178%), tantalum (0.05%), other (0.10%–0.40%) and titanium (in balance).

The machinability tests were performed using a Kingsbury MHP 50 CNC lathe. The tools used were the VNMG 160404-QM H13A insert with the DVJNL 2020K16 tool holder (Fig. 6.1), both from Sandvik Coromant. The geometry of the tool holder is defined by: tool cutting edge angle (93 degrees), inclination angle (−13 degrees), and orthogonal rake angle (−4 degrees). An important effect of the negative inclination was machining with chip flow to left of the "nonmachined side," which can reduce scratches on the machined surface.

The experimental tests were divided into two classes, long and short. The long tests were used to observe the tool life; two parameter combinations were used: (a) cutting speed (vc) of 30 m/min, feed rate (*f*) of 0.1 mm/rev, and depth of cut (a_p) of 0.50 mm; (b) cutting speed (vc) of 90 m/min, feed rate (*f*) of 0.1 mm/rev, and depth of cut (a_p) of 1.00 mm. The short tests were developed to analyze the cutting forces and the surface roughness. In these tests, the cutting speeds were defined as 30 (low speed), 60 (conventional speed), 90 (speed of transition), and 120 m/min (high speed); the feed rate was fixed at 0.05 and 0.10 mm/rev; and depth of cut was at 0.25 and 0.50 mm.

For monitoring the tool wear, a Mitutoyo TM 505 toolmakers' microscope, equipped with a Motic Moticam 2.0 CMOS cam, Motic MLC-150 fiber optic illumination, and Motic Images Plus 2.0 software, was used. The short tests were developed

(A) (B)

Fig. 6.1 Dual negative tool set-up used in the machinability tests. (A) Insert VNMG 160404-QM H13A. (B) Tool holder DVJNL 2020K16.

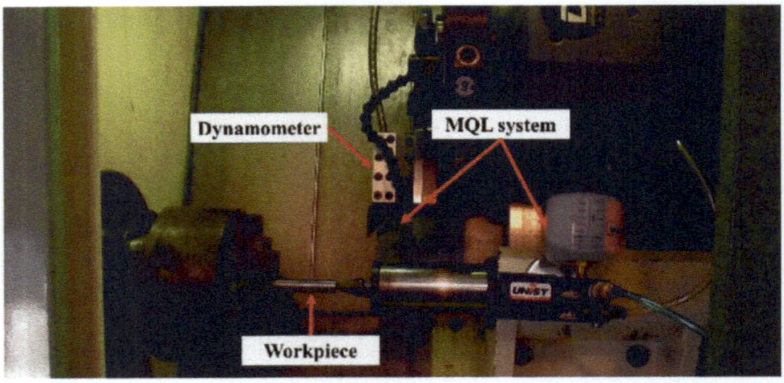

Fig. 6.2 Experimental set-up for the machining tests with dual negative tools.

employing tools with different conditions, both new and worn. The tool condition was evaluated using the flank wear at the corner (V_{Bc}), with the tools considered new when the V_{Bc} was lower than 0.03 mm. In the worn condition, a tool with flank wear (V_{Bc}) between 0.20 and 0.25 mm was used.

A Kistler piezoelectric dynamometer model 9121 connected to a charge amplifier model 5019B and Dynoware software with a computer interface were used to measure the cutting forces. The tests were carried out using dry and cooling/lubricated conditions. For the cooling/lubricated condition, an MQL system composed of a Unist MQL model Coolubricator JR and biodegradable oil Coolube 2210 was employed. The flow rate was fixed at 2.6 ml/min and the pressure at 4 bar. In Fig. 6.2, the experimental set-up for the machinability tests of the Ti-6Al-7Nb alloy is shown.

To analyze the surface roughness, measurements were performed every 120 degrees around the surface of the bars using a Hommel Tester T1000 profilometer and Turbo Datawin software. The workpiece was measured after cleaning at room temperature, and the cut-off was fixed at 0.8 mm to improve the characterization of the surface. The short tests were performed randomly and repeated three times.

6.3 Results and discussion

The tool life is shown in Fig. 6.3. When considering the end of tool life, by employing the MQL system, the tool life is prolonged in 30% of the tests. During the accommodation of the cutting edge, the dry condition presented lower flank wear, but after this stage, the MQL condition provided lower flank wear. In the first moments of tool life, the irregularities of the cutting edge are eliminated [21, 22].

For the analysis of forces in the machinability tests, each component, F_x (depth of cut direction), F_y, (feed direction), and F_z (tangential direction), was analyzed separately. The evolution of the F_x force is shown in Fig. 6.4. The depth of cut was the parameter that caused more variation in the F_x force. When increasing the depth of cut, an average increase of 32.1% in the F_x force was observed. The lowest variation was obtained for the cutting with worn tools and cooling/lubrication, at 12.9%, and the highest for the cutting with new tools and cooling/lubrication, at 51.9%. The

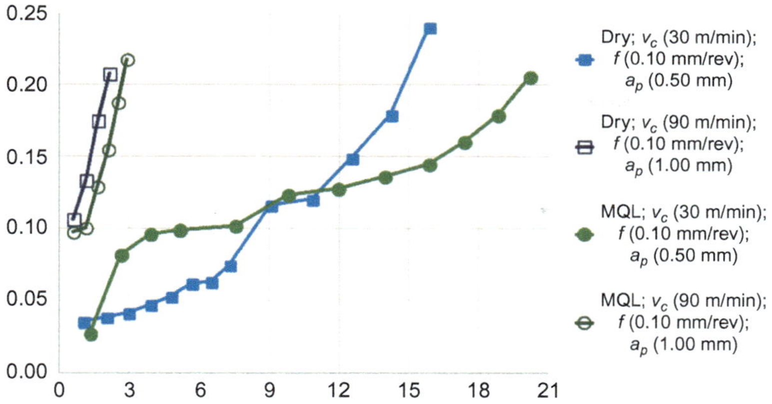

Fig. 6.3 Tool life in the machinability tests.

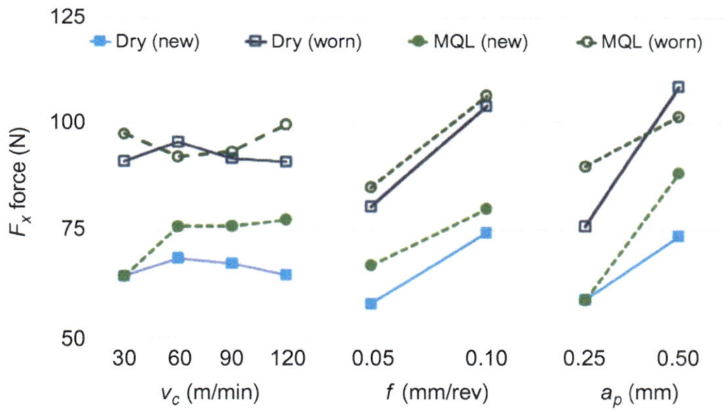

Fig. 6.4 Main effects for F_x force in the machinability tests.

growth of the feed rate caused a higher increase of the F_x force in the dry condition, by 28.1% (new) and 32.7% (worn), than in the MQL condition, at 21.0% (new) and 28.4% (worn).

In general, for the dry cutting the evolution of the F_x force showed a decreasing trend with increase in cutting speed. For the cutting speed, the F_x force increased for the variation from 30 to 60 m/min, at an average of 6.3%, and decreased for other values, at an average of 0.7%. The new and worn tools presented the opposite evolution when using the MQL system. In particular, the new tools reached a peak for 60 m/min, and for the worn tools the lowest value was obtained for 60 m/min.

The main effect for the F_y force is shown in Fig. 6.5. The growth in the feed rate provided an increase in the F_y force for both conditions, at 13.8% (dry) and 15.7% (MQL). When increasing the depth of cut, the F_y force increased steeply, using both dry (153.0%) and MQL (173.6%) conditions. The increase in cutting speed showed a decreasing trend for the F_y force, except for the new tools with MQL. For the combination

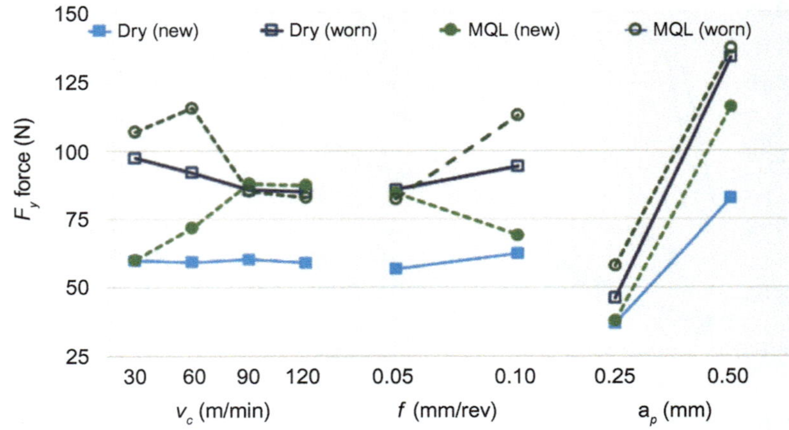

Fig. 6.5 Main effects for F_y force in the machinability tests.

of the MQL system and the new tool, the F_y force decreased when the feed rate was increased for the tests with depth of cut of 0.50 mm. This may have occurred due to the increase in the cutting section improving the stability of the cutting and material flow. Maranhão and Davim [23] studied the relation of the friction coefficient and the material flow. These authors found that an increase in the material flow reduces the friction coefficient. Furthermore, the use of the MQL system improved the cutting because it reduced the friction and facilitated chip pull-off [24]. Leppert [25] observed a growth in the cutting force components when employing low levels of feed rate.

The main effect for the F_z force is shown in Fig. 6.6. The F_z force increased when increasing the depth of cut and feed rate, for all conditions, at 94.0% and 53.4%, respectively. For the cutting speed, a reducing trend of the F_z force was observed when increasing the cutting speed, except for new tools with the MQL system.

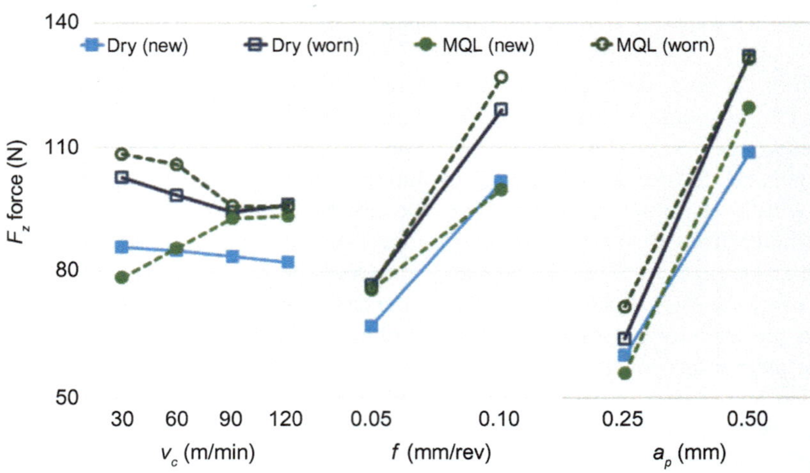

Fig. 6.6 Main effects for F_z force in the machinability tests.

Fig. 6.7 shows the main effect for the R_a (arithmetic average) surface roughness. Analyzing the obtained values of R_a when increasing the cutting speed, the values of R_a diminished for most of the tested conditions. A rise in the feed rate sharply increased R_a in 149.5% (dry) and 173.6% (worn). When increasing the depth of cut from 0.25 to 0.50 mm, the R_a showed a reduction (19.6%, on maximum) or increase (151.8%, on maximum). When considering the growth of the depth of cut for worn tools, the values of R_a decreased. This can also be justified due to the improvement of the stability of the cutting. Employing the cooling/lubrication resulted in a reduction of the values of R_a for most of the conditions, both for new (17.4%) and worn tools (3.2%). A reduction in R_a was observed when using worn tools, 14.7% (dry) and 6.3% (MQL).

The main effects on values of R_t (total height) surface roughness are plotted in Fig. 6.8. For R_t, a similar evolution to that of R_a can be observed. When increasing the feed rate, R_t increased in 37.6% (dry) and 92.9% (MQL). For cutting speed, R_t reached a peak at 60 m/min and the lowest value was obtained for 90 m/min. An increase in the depth of cut caused variations in the R_t surface roughness, with a decrease of 25.1% or an increase of 280.8% occurring, as a maximum. These reductions were observed for worn tools, mainly, when employing the MQL system. The use of the MQL system reduced the values of R_t in 26.5% (new) and 12.4% (worn). When comparing the tool condition, R_t values for worn tools were lower than those for new tools, 14.9% (dry) and 1.5% (MQL), on average.

To analyze the correlation between the variables, the Spearman's ρ correlation parameter was used. The Spearman's ρ is appropriate to analyze correlation between some parameters because it identifies monotonic relationships, is resistant to the effect of outliers, and does not assume a specific parametric model or specific distributions for the data [26].

When analyzing the correlations, shown in Fig. 6.9, it can be observed that the depth of cut and the feed rate showed stronger correlations than the other parameters, which can be explained based on the investigation developed by Fang [17]. The author

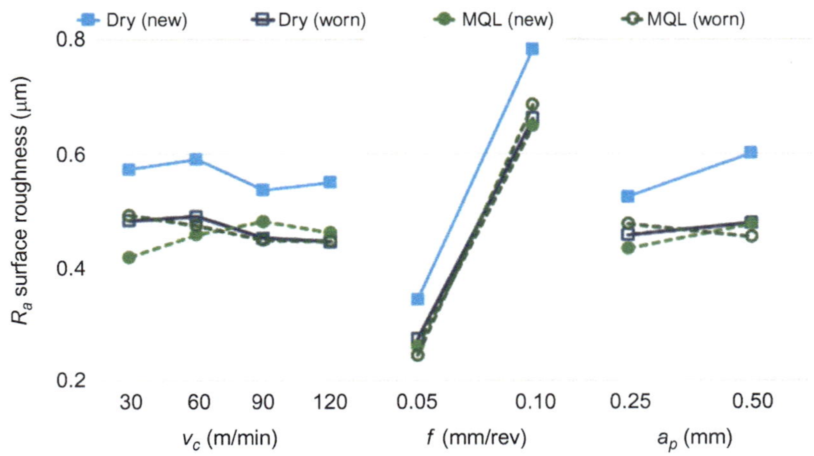

Fig. 6.7 Main effects for R_a surface roughness in the machinability tests.

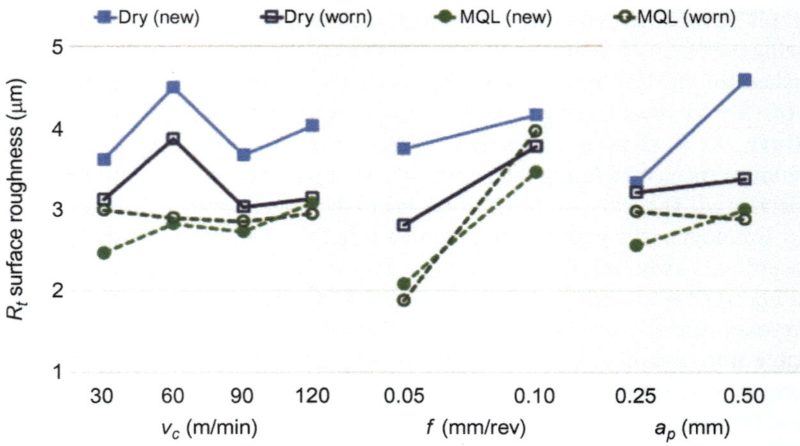

Fig. 6.8 Main effects for R_t surface roughness in the machinability tests.

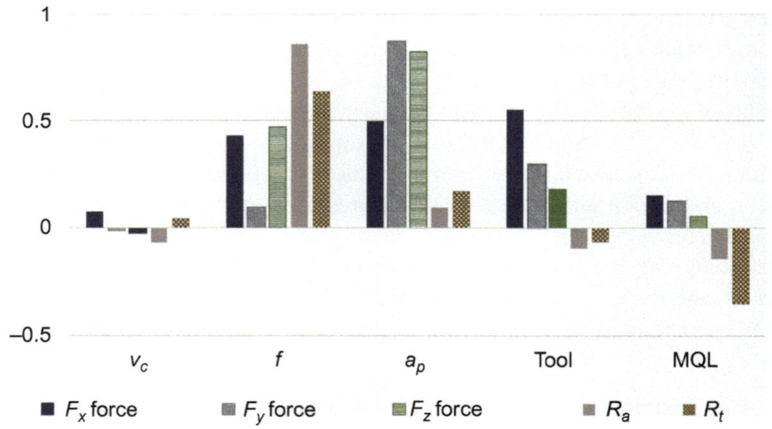

Fig. 6.9 Spearman's ρ correlation for the response for the machinability tests.

analyzed the force and the chip thickness ratios, highlighting that rake angles, both positive and negative, can affect the processes differently. Furthermore, for negative rake angle geometry, an increase in the contact area and machining chip volume provides a considerable growth in the cutting forces and heat generation [27]. Another aspect to consider is the Ploughing effect, which prevailed in the finish hard turning employing a negative rake angle. In addition, the nose radius is large compared to the depth of cut and the feed rate [28]. Thus, when the nose radius is larger than the feed rate, in some cases, tearing of the material can occur instead of the removal of the material by shearing, favoring the Ploughing phenomenon.

The performance of the new tools employing the MQL system might have been influenced by the tool life. Leppert [25] stated that the effect of the cooling/lubrication is related to the cutting parameters employed. The author observed that an increase in the cutting speed decreased the influence of the cooling and lubrication conditions. In addition, effective for values lower than 100 m/min the nozzle position is important, because lower cutting forces can be reached when oil mist is provided to rake and flank faces [29]. Thus, the analyzed geometry of the cutting tools can interfere with the oil mist supplied to the rake and flank faces.

When compared to the theoretical values using Eqs. (6.1), (6.2), the experimental results for the feed rate of 0.05 mm/rev were higher than the $R_{a\text{-}th}$ (0.20 µm) and $R_{t\text{-}th}$ (1.56 µm) values, by 25.2% and 31.5%, respectively, on average. For the feed rate of 0.10 mm/rev, the $R_{a\text{-}th}$ (0.78 µm) and the $R_{t\text{-}th}$ (6.25 µm) were lower than the theoretical values, by 14.4% and 65.7%, respectively, on average. The obtained values of R_a and R_t for the low feed rate can be justified due to the ploughing effect, which can be more noticeable since the relation with the cutting-edge radii was reduced by 50%, and the feed rate offers lower stability in the cutting.

So,

$$R_{a-th} = \frac{f^2}{32 \cdot r_e} \cdot 1000 \tag{6.1}$$

where $R_{a\text{-}th}$ is the theoretical arithmetic average surface roughness (µm); f is the feed rate (mm/rev); and r_e is the nose radius (mm); and

$$R_{t-th} = \frac{f^2}{8 \cdot r_e} \cdot 1000 \tag{6.2}$$

where $R_{t\text{-}th}$ is the theoretical total height surface roughness (µm); f is the feed rate (mm/rev); and r_e is the nose radius (mm).

6.4 Conclusions

Based on the results of the experiments in the turning of the Ti-6Al-7Nb alloy using tools with unconventional geometry, negative inclination/rake angles and a chip breaker, the following conclusions can be drawn:

- An increase in the F_x force was observed when the depth of cut and the feed rate were increased, but it showed no correlation. Among the inputs, the tool wear was the only variable that showed a monotonic correlation with the F_x force.
- The F_y force had a monotonic correlation with the depth of cut. Although the cutting speed did not show correlation, it can be observed that the increase of the cutting speed reduced the feed force in some of the tested conditions.
- For the F_z force, the cutting speed showed no correlation. However, the increase of the cutting speed favored a reduction of the cutting force in some conditions. The depth of cut showed a strong correlation, which practically doubled the values of the force when the depth of cut was increased.
- For the feed rate of 0.05 mm/rev, lower values of R_a and R_t surface roughness than those obtained for the 0.10 mm/rev were observed. However, when compared with the theoretical values, the experimental values for the feed rate of 0.10 mm/rev were lower than the theoretical value, while the experimental values for 0.05 mm/rev were higher than the theoretical value. Furthermore, the MQL system showed favorable results in terms of surface quality.

Acknowledgement

The authors would like to thank TiFast S.r.l. and Sandvik Coromant (Portugal). Additional thanks go to the Coordination for the Improvement of Higher Education Personnel (CAPES), Brazil. The authors acknowledge "Project No. 031556-FCT/02/SAICT/2017; FAMASI—Sustainable and intelligent manufacturing by machining, financed by the Foundation for Science and Technology (FCT), POCI, Portugal.

References

[1] E. Trent, P. Wright, Metal Cutting, fourth ed., Woburn, MA, Butterworth–Heinemann, 2000. 446 p.

[2] M.E. Merchant, An interpretive look at 20th century research on modeling of machining, Mach. Sci. Technol. 2 (2) (1998) 157–163. Available from: http://www.tandfonline.com/doi/abs/10.1080/10940349808945666.

[3] L.E.A. Sanchez, H.J. Mello, R.R.I. Neto, J.P. Davim, Hot turning of a difficult-to-machine steel (sae xev-f) aided by infrared radiation, Int. J. Adv. Manuf. Technol. 73 (5) (2014) 887–898. Available from: http://link.springer.com/10.1007/s00170-014-5879-3.

[4] H. Coldwell, R. Woods, M. Paul, P. Koshy, R. Dewes, D.K. Aspinwall, Rapid machining of hardened AISI H13 and D2 moulds, dies and press tools, J. Mater. Process. Technol. 135 (2–3) (2003) 301–311. Available from: http://linkinghub.elsevier.com/retrieve/pii/S0924013602008610.

[5] R. Ferreira, D. Carou, C.H. Lauro, J.P. Davim, Surface roughness investigation in the hard turning of steel using ceramic tools, Mater. Manuf. Process. 31 (5) (2016) 648–652. Available from: http://www.tandfonline.com/doi/abs/10.1080/10426914.2014.995051?journalCode=lmmp20.

[6] A. Shokrani, V. Dhokia, S.T. Newman, Investigation of the effects of cryogenic machining on surface integrity in CNC end milling of Ti–6Al–4V titanium alloy, J. Manuf. Process. 21 (2016) 172–179.

[7] S.I. Jaffery, M. PTT, Assessment of the machinability of Ti-6Al-4V alloy using the wear map approach, Int. J. Adv. Manuf. Technol. 40 (7–8) (2009) 687–696. Available from: http://link.springer.com/10.1007/s00170-008-1393-9.

[8] L.S. Ahmed, N. Govindaraju, M. Pradeep Kumar, Experimental investigations on cryogenic cooling in the drilling of titanium alloy, Mater. Manuf. Process. 31 (5) (2016) 603–607. Available from: http://tandfonline.com/doi/full/10.1080/10426914.2015.1019127.

[9] Z. Hui, Z. Liu, Z. Cao, M. Qiu, Effect of cryogenic cooling of tool electrode on machining titanium alloy (Ti–6Al–4V) during EDM, Mater. Manuf. Process. 31 (4) (2016) 475–482. Available from: http://tandfonline.com/doi/full/10.1080/10426914.2015.1037893.

[10] J. Gallego, T.S. Pinheiro, R.Z. Valiev, V. Polyakova, C. Bolfarini, C.S. Kiminami, et al., Microstructural characterization of Ti-6Al-7Nb alloy after severe plastic deformation, Mater. Res. 15 (5) (2012) 786–791. Available from: http://www.scielo.br/scielo.php?script=sci_arttext&pid=S1516-14392012000500016&lng=en&nrm=iso&tlng=en.

[11] H. Wu, L. Guo, Machinability of titanium alloy TC21 under orthogonal turning process, Mater. Manuf. Process. 29 (11–12) (2014) 1441–1445. Available from: http://tandfonline.com/doi/full/10.1080/10426914.2014.952023.

[12] M. Geetha, A.K. Singh, R. Asokamani, A.K. Gogia, Ti based biomaterials, the ultimate choice for orthopaedic implants—a review, Prog. Mater. Sci. 54 (3) (2009) 397–425. Available from: http://www.sciencedirect.com/science/article/pii/S0079642508001126.

[13] M.K. Gupta, P.K. Sood, V.S. Sharma, Machining parameters optimization of titanium alloy using response surface methodology and particle swarm optimization under minimum quantity lubrication environment, Mater. Manuf. Process. 31 (13) (2015) 1671–1682. Available from: http://tandfonline.com/doi/full/10.1080/10426914.2015.1117632.
[14] V.P. Astakhov, J.P. Davim, Tools (geometry and material) and tool wear, in: J.P. Davim (Ed.), Machining—Fundamentals and Recent Advances, first ed., Springer London, London, 2008, pp. 29–57. Available from: http://www.springerlink.com/index/10.1007/978-1-84800-213-5.
[15] K. Cheng, Machining Dynamics—Fundamentals, Applications and Practices, Springer, London, 2009328.
[16] S. Ozturk, E. Altan, Slip-line metal cutting model with negative rake angle, J. Braz. Soc. Mech. Sci. Eng. 34 (3) (2012) 246–252. Available from: http://www.scielo.br/scielo.php?script=sci_arttext&pid=S1678-58782012000300004&lng=en&nrm=iso&tlng=en.
[17] N. Fang, Tool-chip friction in machining with a large negative rake angle tool, Wear 258 (5–6) (2005) 890–897. Available from: http://www.sciencedirect.com/science/article/pii/S0043164804003333.
[18] Sandvik Coromant, General turning (Positive and negative). Knowledge, Available from: http://www.sandvik.coromant.com/en-gb/knowledge/general_turning/how-to-achieve-good-component-quality/insert-information/positive-and-negative/Pages/default.aspx, 2015.
[19] K.C. Ee, A.K. Balaji, I.S. Jawahir, Progressive tool-wear mechanisms and their effects on chip-curl/chip-form in machining with grooved tools: an extended application of the equivalent toolface (ET) model, Wear 255 (7–12) (2003) 1404–1413. Available from: http://www.sciencedirect.com/science/article/pii/S0043164803001121.
[20] A. Maurotto, R. Muhammad, A. Roy, V.V. Silberschmidt, Enhanced ultrasonically assisted turning of a β-titanium alloy, Ultrasonics 53 (7) (2013) 1242–1250. Available from: http://www.sciencedirect.com/science/article/pii/S0041624X13000693.
[21] J.A. Arsecularatne, L.C. Zhang, C. Montross, Wear and tool life of tungsten carbide, PCBN and PCD cutting tools, Int. J. Mach. Tools Manuf. 46 (5) (2006) 482–491. Available from: http://www.sciencedirect.com/science/article/pii/S0890695505001537.
[22] G. Poulachon, A. Moisan, I.S. Jawahir, Tool-wear mechanisms in hard turning with polycrystalline cubic boron nitride tools, Wear 250 (1–12) (2001) 576–586. Available from: http://www.sciencedirect.com/science/article/pii/S0043164801006093.
[23] C. Maranhão, J.P. Davim, The role of flow stress and friction coefficient in fem analysis of machining: a review, Rev. Adv. Mater. Sci. 30 (2) (2012) 184–188.
[24] D. Carou, E.M. Rubio, J.P. Davim, A note on the use of the minimum quantity lubrication (MQL) system in turning, Ind. Lubr. Tribol. 67 (3) (2015) 256–261. Available from: http://www.emeraldinsight.com/doi/full/10.1108/ILT-07-2014-0070.
[25] T. Leppert, Effect of cooling and lubrication conditions on surface topography and turning process of C45 steel, Int. J. Mach. Tools Manuf. 51 (2) (2011) 120–126. Available from: http://www.sciencedirect.com/science/article/pii/S0890695510001902.
[26] D. Carou, E.M. Rubio, C.H. Lauro, J.P. Davim, Experimental investigation on finish intermittent turning of UNS M11917 magnesium alloy under dry machining, Int. J. Adv. Manuf. Technol. 75 (2014) 1417–1429.
[27] M. Günay, İ. Korkut, E. Aslan, U. Şeker, Experimental investigation of the effect of cutting tool rake angle on main cutting force, J. Mater. Process. Technol. 166 (1) (2005) 44–49. Available from: http://www.sciencedirect.com/science/article/pii/S092401360400994X.

[28] Y. Huang, S.Y. Liang, Force modelling in shallow cuts with large negative rake angle and large nose radius tools—application to hard turning, Int. J. Adv. Manuf. Technol. 22 (9–10) (2003) 626–632. Available from: http://link.springer.com/10.1007/s00170-003-1550-0.
[29] M. Hadad, B. Sadeghi, Minimum quantity lubrication-MQL turning of AISI 4140 steel alloy, J. Clean. Prod. 54 (2013) 332–343. Available from: http://linkinghub.elsevier.com/retrieve/pii/S0959652613003181.

Index

Note: Page numbers followed by *f* indicate figures and *t* indicate tables.

T